The New Experimental Design

Teruo Mori

Originally published in Japanese under the title of "Taguchi Mesoddo o Tsukatta Yasashi — Shin Jikken Keikakuho Nyumon", by Torendo Books, Japan. Copyright 1987, by Teruo Mori.

This is an authorized English translation from the Japanese edition under agreement with Teruo Mori.

A number of other titles are available from the ASI Press including:

The Customer-Driven Company: Managerial Perspectives on QFD

Quality By Design: Taguchi Methods and U.S. Industry

Taguchi Methods: Case Studies from the U.S. and Europe, Quality Engineering Series, Volume 6

Taguchi Methods and QFD: Hows and Whys for Management

The New Experimental Design

Taguchi's Approach to Quality Engineering

Teruo Mori

ASI Press

Published by the ASI Press, an ASI company. Printed in the United States of America.

Library of Congress Cataloging-in-Publication Data

Mori, Teruo, 1946–
[Taguchi mesoddo o tsukatta yasashii shinjikken keikakuhō nyūmon. English]
The new experimental design : Taguchi's approach to quality engineering / Teruo Mori.
p. cm.
Translation of: Taguchi mesoddo o tsukatta yasashii shinjikken keikakuhō nyūmon.
Includes index.
ISBN 0-941243-13-3 : $45.00
1. Taguchi methods (Quality control) 2. Experimental design. I. Title
TS156.M64713 1991 90-24031
658.5'62--dc20 CIP

Contents

INTRODUCTION AND RECOMMENDATION

The author, Mr. Teruo Mori, is a key member of the Quality Control Research Group of the Japanese Standards Association. He previously authored *Case Studies in Experimental Design*, which was the type of book that I had wanted to see brought out for many years.

Mori has revised and rewritten that earlier book into what one could call a new publication. Whereas the former book had more volume, this book is compact and constitutes a new, improved edition in many aspects.

Experimental design itself consists of only two techniques:

1. Layout of factors (assigning many factors to an orthogonal array)
2. Analysis of variance (ANOVA).

It is, indeed, most important for us to know where to apply these techniques in the various technological fields.

When using experimental design in the design of new products or the development of a manufacturing process, the application of analysis of variance, to obtain a reasonable estimate of functional quality is of foremost importance. This includes the following three items;

a. Selection of characteristics for measurement
b. Selection of test conditions
c. Selection of the appropriate Signal-to-Noise (S/N) Ratio as an estimator characteristic for determining optimum conditions.

These points require individual solutions for each and every problem encountered in development and design research. That may be the reason that Mori's book is titled *New Experimental Design in Japan*, and avoids terminology of the past, in order to explain this individualistic method of application.

Mori realizes the extreme importance of the three items a., b. and c. in the rationalization of development and design research Chapter 3 describes the many methods of expressing functional characteristics. In order to apply these to particular problems, I recommend that readers first master the contents of Chapter 3 and then proceed to the case studies outlined in Chapter 6. Countermeasures (a.), (b.) (c.), are based on technological insight of the product's essential functions. These and an understanding of the techniques of analysis of variance provide the sole method for rationally solving the problem of characteristic values.

In his previous book, Mori devoted many pages to the following:

1. The layout of complex factors including interactions
2. ANOVA techniques

In his new book, however, he devotes fewer pages to significance testing and estimation, which I feel is a good thing. He writes in more detail on the meaning of appropriate design and how to find an optimum design and optimum conditions. He also gives advice for dealing with the various problems that may arise.

I think that Mori's use of the L_{18}, L_{12}, and other orthogonal arrays in place of L_{16} and L_{27} arrays is a particularly good point. The S/N Ratio, a measure of a function's stability, is itself an estimator measurement that deals with the interaction between controllable and noise variables. Accordingly, interactions between controllable factors are secondary interactions. I agree with Mori's recommendation of using an L_{12}, L_{18}, and L_{36} arrays and not analyzing interactions. I think this is a very important point. To understand the purpose of using L_{18} and L_{12} arrays, see Chapter 2.

For analysis of the data and case studies in this book, Mori has developed and utilized a powerful program for personal computers. There have been many programs developed for analysis of experimental designs, particularly programs using S/N ratios. The majority of such programs lack flexibility, but the program introduced in this book has wide application.

In the world of technology and design, creation of a product having low cost and ideal functions is of utmost importance. Chapter 1 emphasizes this. In tolerance design, improvements in functional quality can also be used to reduce cost. The issue of finding a balance between quality and cost is not fully explained in this book. However, as an introductory work, this may be unavoidable. I recommend this book, without reservation, to the many people in charge of research and development and design.

DR. GENICHI TAGUCHI

PREFACE

On several occasions I have had the opportunity to discuss corporate competition with American business executives and managers. The main topic of discussion was usually the ranking of international competition among countries. When Japan was given 10 points, the following comparison eventually emerged:

Grade	Country	Performance
Best	Japan	10
Better	West Germany U.S. France Korea	5 – 6
Good	China	2

Business executives would say the reason Japan gets the highest score is because of its products' life and low price relative to high quality.

The next topic was how to make a higher-quality product less expensively, as the Japanese do. Of course, Americans have been studying the differences between Japan and the United States for more than a decade and most of them maintained that the following areas are where Japan is superior to the USA:

1. Operators (workers)
   Japanese workers have an interest in how their operation affects the production and profit of the company, so they accept daily management activities, such as Quality Circles.
2. Engineers
   Japanese engineers always keep product cost and quality as defined by the consumer in mind. Thus, the resulting product is of low cost and high quality.
3. Managers
   Japanese managers maintain a long term perspective and, therefore, have an interest in improving company-wide management quality for the future.

On the other hand, the USA is superior to Japan in the following respects:

1. Development of new key technologies
   Most new key technologies, such as superconductivity, biotechnology, and computers, have originated in the United States.
2. Development of large-scale industries
   Technologies and industries using large-scale facilities and equipment, such as the nuclear power, space and marine industries, etc., have been developed in the United States.
3. Availability of many excellent scientists
   The United States has many excellent scientists, including numerous Nobel Prize recipients.

According to the business executive's conclusions, Japan is good at developing product technology and the United States is good at developing new key technologies.

If the United States is to make cheaper and higher quality products than Japan, what should it do? Small group activities and Total Quality Control from Japan will provide good effects toward this end. But this will not be enough to catch up with Japanese cost and quality. Simply imitating Japanese management style will not put the United States ahead of Japan. At minimum, a product must satisfy the following points at the development stage:

1. It should be of the lowest cost possible at all stages from raw material to the consumer. (Low cost.)
2. It should satisfy its target values. (Stable quality.)
3. Its development time should be quicker than that of its competitors. (Short development time.)

These points constitute the ideal conditions for new product development, manufacturing equipment, and optimum process operation. We need new philosophies and concrete methods to accomplish these goals. The purpose of this book is to introduce a clear philosophy and detailed methods for realizing these ideal development conditions in order to develop competitive new products at the design stage.

Dr. Genichi Taguchi has developed goals and introduced a reasonable philosophy and concrete method for meeting these goals. His system is called "Quality Engineering." It is also known as "Taguchi Method," "Robust Design," and other names throughout the world. Its basic principle is simple, but one must understand many difficult points in trying to grasp it. I have divided the system's basic philosophy and methodology into reasonable parts in order to make it easier for readers who work in the design field.

The main contents are as follows:

1. A difference in design philosophy between the traditional and the new approach (Chapters 1 - 3)

   This part of the book is written in a conversational style and clarifies the differences between the two approaches. You will discover the strong points and special merits of the new approach relative to the traditional philosophy.

2. Manual of design steps (Chapters 4 - 5)

   I have compiled a standard manual of concrete methods for making achievement of the ideal conditions easy. This will clear up any confusion about the development stage. Because the manual includes all types of procedures and terms relating to design, it will transform new engineers into experts.

3. Case studies (Chapters 6 - 9)

   This book has many case studies that follow the manual (Chapters 4–5). The cases are picked from various industries, such as the chemical, mechanical, metalworking, and electronics industries. The cases have been selected to give the reader clear and reasonable suggestions for dealing with the many kinds of problems encountered at the design stage. They will help make any type of design successful.

4. Computer analysis programs (Chapter 10)
I have described some useful personal computer programs. Readers can easily adapt and develop them. They make complex calculations easy and also provide additional information.

I hope that many people will make use of this book; that it will bring out untapped abilities of engineers, designers, and researchers; and that it will help them in their careers.

I wish to express my gratitude to Dr. Taguchi and the members of Quality Control Research Group for making the publication of this book possible. I also wish to thank Chairman Yoshiaki Abe of the Business System Research Institute and President Sasaki of Trend Books.

TERUO MORI

1

Importance of Experimental Design Today

1-1. Corporate Competition in the Multi-Product Small Quantity Production Era

Q. What are the characteristics of corporate competition in an era of small-quantity production of many kinds of products?

M. The following three factors are fast creating a diversification of products in the market today:

1. Rapid technological innovation
2. Shortened product life cycle
3. Diversification of consumer needs.

Table 1-1 summarizes the factors causing product diversification. It separates them into consumer and producer factors.

Table 1-1: Factors Creating Product Diversification

	Consumer-Side Factors	Producer-Side Factors
1.	Demand for sophistication and value diversification with increased income	New-product development to expand demand
2.	Fine market segmentation based on selective demand caused by diversification in values	Increased new product lines based on a development philosophy that lacks coherence
3.		More sophisticated and complicated technology and systems Increase in machine types resulting from the composition of existing independent technologies

This trend toward multi-product, small-quantity production runs counter to traditional corporate activities, such as cost reduction and improvements in production efficiency, as well as research and development. It increases cost in every area and tremendously reduces corporate competitiveness.

Q. Several methods are applied in the area of material and production management. These include unification in design ideas through the use of common units and part series, reduction in die-setting time, and the production of many kinds of products on a standard line. However, the design department still appears to be in an era of few-product mass production.

M. *Table 1-2* compares the differences in development of few-product mass production and multi-product, small-quantity production.

Table 1-2: Comparison of the Development of Few-Product Mass Production with Multi-Product, Small-Quantity Production

	Few-product, Mass Production	**Multi-product, Small-Quantity Production**
Key section	Production section	Design department
Mass productivity	The production section performs activities to reduce the defect rate over a long time period	The design department determines the optimum condition, including the cause of variability
Cost down	Cost down is achieved by the mass-production effect based on the QC activities within the production section	Cost reduction within the production section is impossible with small-quantity production
Development Period	Long time	Short time

With multi-product, small-quantity production, **the key department has shifted from the production department to the design department.** In the mass-production era, production departments were supplied with insufficiently executed designs, which generated products with many defects. It was assumed that the on-line department would carry out activities to achieve zero defects over a long period of time. The members of the production department were proud of their achievements in reducing the defect rate of the high defect-generating designs initially given to them.

Q. In the era of multi-product, small-quantity production the product must be free of defects from the start of production. This is true despite the fact that the development time is shortened and fewer researchers are available because there are numerous research themes. However, it is impossible to quickly and inexpensively create a **robust** (stable and reliable) product in the design stage.

M. The fact is **not** that you cannot execute such a design. Rather, we have not had a design method that makes it possible to achieve this objective. We need to change our way of thinking regarding design.

I recommend the use of an experimental design method supported by a new philosophy and based on new concepts. Such a method has been developed by Dr. Genichi Taguchi. It is called Quality Engineering (or Taguchi Methods in the United States) and combines engineering and statistical methods to achieve improvements in cost and quality by optimizing product design and manufacturing processes.

Table 1-3 compares the methods used to determine optimum development conditions. They are the **multi-factor layout**, the **one-factor-at-a-time experiment**, the **target method**, the **traditional experimental design method**, and the **Quality Engineering method** (Taguchi Methods).

This table is based on results from my personal experience. I would like readers to judge their own activities based on the evaluation format in *Table 1–3*.

Q. The traditional method places priority on mathematical analysis, but what about the new method?

M. There certainly is a method that uses many mathematical and statistical analyses. **But it is most important that the engineer comes up with a robust design quickly and inexpensively.** The best way to achieve this objective is to learn and apply the Quality Engineering method (Taguchi Methods). Survival of a corporation depends on its philosophically based development activities, together with a method equipped with concrete concepts.

1-2. The Role of Design in Corporate Profit

Q. Is the process of achieving an optimum solution different for different types of researchers?

M. Research can be classified into two categories: design-oriented and analysis-oriented. The approach used to obtain the optimum solution for each category is different. **The objective of the design-oriented researcher is to offer an optimum solution to an engineering objective. The objective of the analysis-oriented researcher, on the other hand, is to create a model that expresses the system.**

Q. How does the process of obtaining the optimum solution differ between design and analysis?

M. Because the objective of the design-oriented researcher is to find the optimum solution to research problems, such as new-product development, equipment design, and process design, we must consider costs, the development period, and product stability.

Design is the only action that materializes corporate activity. Therefore, priority is placed on quickly getting good, inexpensive products from the production floor to the market.

Table 1-3: Comparison of Methods for Determining the Optimal Condition

Item	Content	Method Used to Obtain the Optimal Condition: Multi-Factor	One Factor at a Time	Target Method	Traditional Design of Experiments	Taguchi Methods
No. of experiments (development burden)	Can a small no. of experiments suffice?	81* Many	12* Relatively small	2–3* Quite small	27* Moderate	9* Small
Production cost	Is it possible to produce an inexpensive product?	Impossible	Impossible	Impossible	Impossible	Possible
Robust design (stability and reliability of the optimum condition)	Are the data reproducible?	Good	None	None	Moderate	Best
	Can it cope with changes in exogenous factors?	Moderate	Weak	Weak	Weak	Strong
	Can prototype data be applied to production model?	Impossible	Impossible	Impossible	Impossible	Possible
Comments on the method		Impractical, no. of experiments is too large.	Optimum condition lacks reliability. There is no reproducibility.		Uniform allocation of error factor and significance test conducted. Includes interaction and error terms. Complicated calculations are not practical.	With new philosophy, orthogonal array is effective in optimizing product and process.
Utilization by engineers 90**		0%	53%	45%	0%	0%

* The number of experiments necessary to obtain the optimum condition for four factors with three levels.
** Result from an actual company survey.

Because it relates to every item of product quality and reliability, **the design should hold the most important role in the control of corporate profits.**

Q. Engineers used to obtain the optimum condition in development activities using the one-factor-at-a-time (or the precision test) method. Aren't these methods still applicable?

M. Currently, the major goal in the design stage is to discover a low-cost and robust (stable and highly reliable) new product and new process in a short period of time. In order to achieve this objective, the traditional, analytical-type approaches used in corporations, which include the multi-factor layout, simulation, traditional experimental design methods, and the one-factor-at-a-time experiment, are not suitable.

Most engineers have experienced **the problem of many defects and market claims** when the optimum condition obtained from these methods was applied to the actual product. **This is due to the lack of stability caused by large variability.**

It shows that corporate engineers are using imperfect approaches and methods. When these analytical approaches are used, it is often observed that the more research the person in charge of the design conducts, the more expensive the product becomes or the more sophisticated the parts become. This results in a significantly high cost.

It may come as a surprise to many that the more research one does, the higher the cost of the product. All engineers have had the experience of creating considerable corporate loss because an optimum condition determined in the laboratory or from small-scale equipment was not reproducible. Their laboratory data was completely off from the expected value for the **on-line** process (production machine).

When the optimum condition is off the target, it takes an unnecessarily long development period to adjust the condition. This becomes an important factor in increasing the **defect rate** in the production stage.

Table 1-4 summarizes the traditional analysis method and the results we currently expect from new-product development.

Q. **Engineers realize that the analytical approach they use may be quite far from ideal.** But they do not have a method for bridging the gap. Such a method has not yet been offered to them.

M. That is true. To alleviate this situation, we must provide engineers with a method for obtaining an optimum solution that meets corporate needs (i.e., a method that can be applied at the design stage and results in a new product with low cost, high quality, and stability).

Many factors exist that interfere with effective research and development. *Table 1-5* summarizes such factors. I expect research-and-development engineers to have sufficient management ability regarding development activities and the ability to make adjustments similar to that of management personnel.

Table 1.4. Difference Between the Expected Ideal and Actual in the One-Factor-at-a-Time Experiment in Industry

Item	Content	Ideal	Actual	Comments on the Current Situation
Cost	Cost (equipment, new product)	cheap	expensive	(1)
	Development equipment cost	low	high	(2)
	Production process defect rate	zero	high	(3)
	Market claims	zero	high	(4)
Development period	Optimization at the experimental level	short	long	(5)
	Optimization at the production level	once	several times	(6)
Stability	Production process stability	high	low	(7)
	Product reliability	high	low	(8)

(1) A high-class product and high price is thought to lead to high quality. The system is a combination of high-priced part units.
(2) Experiment equipment similar to the actual production machine is necessary.
(3) The defect rate is high because the optimum condition for the experimental machine does not apply to the production machine.
The design is not robust and changes in the environment and materials increase the defect rate.
(4) A high market claim rate results because the design is not robust and does not allow for use under different conditions.
(5) Reproducibility is low because of the one-factor-at-a-time approach. Thus, it is difficult to determine the optimum condition.
(6) Repeated experiments are necessary because the optimum condition for the experimental machine does not coincide with that of the production machine. This extends the development time.
(7) Stability is low because the design allows for changes in raw material, parts, production environment, and process conditions. This directly influences product performance.
(8) The design is vulnerable to changes in usage conditions and the natural environment because emphasis was placed on setting the mean value of the variables directly related to performance.

Table 1-5. Factors Preventing Effective Research and Development and Countermeasures

The Problem	The Situation	Countermeasures (Approach)
1. The theory is highly regarded, but does not lead to new discovery.	It creates no breakthrough leading to a new product. The approach does not go beyond common sense.	The theory is a mathematical description of reality that describes a simple phenomenon. Invention comes from utilizing complex phenomena. There is no theory for complex phenomena.
2. It is difficult to exercise leadership for a project team of members from different sections.	Development activity is not controlled. Each engineer places priority on his own activity.	The role of the project leader is to assist other members. It is not to control, restrict, or manage. He must: • Help remove obstacles each engineer faces. • Consider the major flow of activity rather than concentrate on detailed management items.
3. Personal ability is not promoted.	The ability of engineers must be developed throughout the project. But this does not occur.	Create good corporate employees rather than just engineers. • Teach them the implications of corporate research and competition. • Instruct them through on-the-job training.
4. Emphasis is on the result. The process is not evaluated.	Good results are praised, but there are not many good results.	Separate evaluation of results from technological capability. Pay attention to even unsuccessful results to boost enthusiasm. • Reward unlimited imagination. • Remove obstacles facing engineers.
5. Attempts are made to remove already known factors.	A solution is not forthcoming because new factors are not picked up.	The objective is to improve the current condition. Results will be unsatisfactory if attention is limited to the factors at hand and factors leading to a systematic solution are not included.

6. The engineer does not have a good understanding of the target or the objective.	The target is considered fixed and absolute.	Corporate activity changes so targets may change. Targets change due to competition with other companies.
7. The experimental design is not well planned. Experimental results are not anticipated.	Experiments are investigations of unknowns. Results cannot be anticipated.	In order to anticipate and remove areas that may lead to failure, ask "If this experiment fails ..." Include recovery measures in the experiment. Check factors, levels, and equipment in the experiment.
8. Theory and mechanism is excessively pursued.	Development is delayed because the engineer pursues a theory and mechanism that will explain the process.	Establish three hypotheses. Examine factors based on these hypotheses to determine which is best in terms of experimental results. Try not to experiment for analytical purposes. If analysis is suggested, use expected results as factors in the experiment.
9. The target is subject to change. This makes it difficult to concentrate efforts.	An undefined target leads to research that is off the point. This is inefficient.	Be aware that targets are *not* included in the standard. The target is a new technology. Development works as a rough guide. The target should be specified as a range of the target variable.
10. Testing cannot be done without equipment.	Research is delayed due to the time required for new equipment to be constructed.	Place priority on constructing handmade, small-sized batch test machines. • Make test machines functional and easy to modify. • Use design activity effectively — determine the design specification from a test batch.

11. Experiments are conducted without expectation of success.	Experiments look similar with slightly different levels of already known factors.	Test design must be re-examined until convinced of achieving the target. Gather the knowledge of every team member. A wild experiment designed only by the chief engineer increases failure rate and leads to chaos.
12. Considerations of cost, productivity, and processing are missing.	Production equipment is difficult to use because quality receives major attention, while cost and productivity are neglected.	Though quality is important, market competition is based on utility = function / cost. Cost must be examined *prior to* quality. Industrial engineering input and cooperation from other sections must be obtained.
13. Attitude toward and evaluation of the experiment are not rigorous.	An attitude of waiting for others exists — "Someone will help us."	Separate facts and estimates. Examine experimental content. Assign various engineers to experimental planning. Beware of information and results that appear too attractive.
14. A good part of the evaluation is based on alternative characteristics.	Real characteristics and alternative characteristics do not necessarily coincide. This causes problems in the final stage.	Important judgments *must be* based on evaluation of real characteristics. (Use alternative characteristics for rough judgments only.) Strength and duration tests are effective in reducing research time.
15. There is error in the perception of the scope of one's responsibility.	There exists a childish attitude in trying to solve the problem with only one's limited ability without asking for help. Attempts are made to solve a problem without asking for help. This is a childish attitude.	Everyone's knowledge and input must be used by the project leader. He should beware of those engineers who perform experiments thinking they know all the answers.

16. Some engineers are obsessed with carrying out experiments rather than addressing the problem.	There are many failures and off-target actions.	Concentrated effort is important, but "There is no wisdom in sweat." Wisdom must come before sweat. Many problems can be solved with wisdom before knowledge.
17. Some managers place priority on the delivery time.	The manager imposes his time limit and does not support his engineers.	The role of the manager is to remove obstacles. An engineer in a senior position must be concerned with the smooth operation of the entire team. Engineers should not be influenced by the personality of the manager. They should take a positive approach rather than just complaining.

1-3. The Technological Capability of Corporate Engineers

Q. Which is more important in the design stage, quality or cost?

M. **I think the cost is more important than the quality.** This is because it is assumed that the quality target will be fulfilled in the prototype stage or in the mass-production stage. Traditionally, priority is placed on developing quality in the development and design stage. Cost is determined on the process without rigorous examination. But it is better to perform research in order to fulfill the target quality at a cost equal to or less than the cost constraint already given. This is true because the lower the cost, the greater the product's contribution to the corporation.

When users decide to purchase a product, the decision is based on its perceived utility. The final product price (sales price) is determined without regard to the standard price suggested by the corporation.

$$\text{Utility} = \frac{\text{Function}}{\text{Price}}$$

The market price of a product is determined through market competition, but the **cost is dependent only on the technological capability of the corporation's engineers.**

The difference between the product's price and its cost is the basic corporate profit.

Q. This means that the engineers' technological capability to produce an inexpensive product is the source of the profit.

M. Yes, indeed. **The price of the product is determined by market competition, regardless of expectations on the part of the corporation.** The cost is controlled only by the corporation's technological capability.

Q. What about the method of first achieving the target quality of the mass-produced product using high-class parts and then attempting to reduce the cost?

M. It is quite difficult to change specifications after the product is mass produced with high-class parts, and any cost reduction achieved will be insufficient. Current competition is based on multi-product, small-quantity production. Modification of the existing product is inefficient, with insignificant effects compared to the huge efforts necessary to achieve it.

I believe that cost reduction should not be treated as a theme in corporate research for two reasons:

1. Priority must be placed on development research of new themes. Research for improving an existing product that will not create significant results must not be pursued in multi-product, small-quantity production.
2. Improvement of an existing product is nothing but an attempt to reduce the cost, which was not accomplished in the design stage. Thus, it is simply a cover-up for mistakes committed in the design stage.

These things must be done in the earliest possible stages. They should not be done after production has begun. Generally speaking, many designs exist which fulfill the quality target, but cost minimization was not considered.

Furthermore, regardless of the cost, there are many cases where the actual development time exceeds what was originally determined to be a reasonable time period.

Q. We have many problems with the cost and the development period. How can we achieve production of low-cost mass-produced products, especially in a short development period?

M. **One must use inexpensive parts and an inexpensive system from the beginning of the research. Technological capability is the ability to derive high performance from low-cost parts and a simple mechanism.**

An average engineer can produce a high-cost product with high-class parts. The worst engineers are the ones who end up with a low-performance product that does not meet the target despite the use of high-class parts. This occurs largely because they lack knowledge about a method for effectively finding the optimum condition.

In order to reduce development time, engineers must study and apply a new philosophy and approach to experimental design.

1-4. The Difference between Science and Engineering

1-4-1. Introduction

There is a critical difference between science and engineering that is difficult to explain in just a few words.

It might be said that the objective of science is to analyze and generalize, while the objective of engineering is to combine and specialize. This section summarizes a conversation I had with some engineers when visiting China. The technical discussion from "Chinese Technology; A Tour Report" is included here to point out some of the differences between science and engineering.

1-4-2. Science and Engineering

This technical section is cast in the form of a discussion between a consulting engineer and his scientist friend who works as an assistant at a university.

The approach, attitude, and philosophy of engineering are presented by comparing a scientist and an engineer. The discussion includes such topics as the importance of process design, what makes bad equipment, and a method for creating a high-quality product from raw materials. The importance of importing not only scientific and technological achievements, but also the approach, attitude, and philosophy toward process equipment design, production, technology, and inspection are emphasized. The discussion should serve as a recommendation for Chinese engineers, as well as a warning to Japanese engineers.

In the following dialogue, "S" indicates the scientist and "E" indicates the engineer.

S. How was your Chinese tour?

E. This trip to China revised my thinking in the following three aspects:

1. I thought China was a poor country, but I was wrong. A great amount of liveliness exists, and the people there live with a rich philosophy. I learned again that material wealth is not the only important thing in life.
2. I had a gloomy image of Socialist and Communist countries, but the people are freer than I expected, as far as the life of the general public is concerned. I think they are succeeding in the development of their country.
3. The level of science is quite high and engineers are studying very hard. We encountered many things that were new to us. However, the level of engineering remains low.

I think these three points summarize the changes in my thinking.

S. "Seeing is believing" applies to the first two points, and only those people who have actually seen China can tell. It sounds like you had a valuable experience. I envy you. By the way, what did you mean when you mentioned in your third point that the level of science is high but the level of engineering is low? Could you please explain that to me?

E. China has several splendid achievements in advanced scientific technology, such as missiles, the atomic bomb, and satellites. However, products produced in their factories cannot compete in the free market, although they are acceptable for practical use.

S. Can you give some examples of what you found in China?

E. For example, there was this aluminum pan. It had streaks (vein-shaped scratches) in it that reduced its competitiveness in the free market, although it can still be used as a pan. The same thing can be said about the merchandise displayed in the Shanghai First Department Store.

S. What is a free market?

E. A free market is a market under a free economy where customers can freely chose and decide to purchase the merchandise available. We can find an unlimited number of examples where a product cannot be sold no matter how sufficiently the product functions or however cheap. A product becomes successful merchandise only when it is competitive.

S. I agree with you. However, I have heard that product types and quantity of production are assigned under the controlled economy in Socialist countries.

E. I don't have a clear idea about product types and quantity of production decisions. But it is true that necessary products in a sufficient quantity are produced.

S. That means that there is no waste, doesn't it?

E. Well, that is not exactly right. It may mean that people are forced to use poor-quality products since there is no room for selection on the part of users. If the product is inferior, it amounts to a loss. In the case of China, the state itself suffers the loss. In Japan, no one will buy the products of a manufacturer who produces inferior products. Such a company goes bankrupt. Japanese manufacturers try hard to avoid such situations.

S. How can we identify inferior products?

E. In the free market, they are the ones that are left unsold.

S. How does that happen?

E. **Customers are wise, and if the product is a disappointment the first time and caused them a loss, they will never buy the same product again.** Those products will become leftovers.

S. Now I understand. How can you identify an inferior product in China?

E. It is difficult to point that out, but I think I can call the product with the lowest utility an inferior product.

$$\text{Utility} = \frac{\text{Function}}{\text{Price}}$$

Function in this case also includes the time from purchase to a breakdown in function (reliability). Cost is the sum of the initial cost of production and the running cost during usage.

S. This means that production of a high-utility product is the production of "a product with minimum cost and maximum function," doesn't it?

E. That is exactly right.

S. There are different models in one car line, such as the ones called SL and DX in Japan. What about in China?

E. That is a phenomenon that only occurs when the merchandise develops under a free market economy.

When the preference of the user diversifies, "differentiation" emerges. In earlier days, Japanese passenger cars consisted of only the Nissan Bluebird and Cedric and the Toyota Corona and Crown. Now it is impossible to remember the names of all of the cars available.

Going back to China, there are not many different models of any one product. I found many kinds of cloth and ceramics, but there was no room for choice in manufactured products. This may be a natural outcome of a controlled economy.

S. I can understand that the economic structure and merchandise quality are related, but what does it mean when you say the level of engineering is low? Could you explain?

E. It means that there are no high-quality products. This is sure proof of a low level of engineering. And products are not competitive without high quality.

S. Is it necessary for the Chinese to have high-quality products?

E. Exporting processed products is more beneficial than exporting raw material because of the higher value added. Because China has a sufficient labor force thanks to its large population, the potential for processing the raw material is there. If a country exports cheap raw materials and imports expensive processed goods, its balance-of-payments position greatly deteriorates.

S. Yes, that explains why newly industrializing countries and developing countries try to improve their technological capability. Now, can you explain how to make high-quality products that are competitive?

E. Before that I would like to tell you of my impression of Chinese engineers.

Our meeting was set up so that the Chinese engineers' questions were handed to us prior to the meeting. We provided our answers during the meeting. I was impressed by the following things:

1. Chinese engineers study very hard. They are well-informed, using academic papers, lectures, and guide books.
2. When we were explaining the details of a production experiment, they kept asking detailed questions, such as what was the temperature, density, formulation, or additives. This pattern was repeated every time we explained conditions.

I felt that Chinese engineers were very enthusiastic. The conversation heated up when they kept asking these questions. It seemed as if they intended to apply the conditions we explained or values indicated in the articles as the factory production condition without any modification.

All they asked was "How to?" They never asked "Why?" or "How much?" questions. It was strange that they never asked the following kinds of questions:

- What was the major factor that determined the condition?
- What were the characteristics of the condition and their shortcomings?
- What was the price?
- What was the approach taken to derive the structure of the formula?
- What were the assumptions of process design?

We think that this kind of meeting is a contact point for small technology transfer.

S. I support that idea.

E. Technology transfer is the transfer of a system that creates good things. I want to limit our attention to the technology here and leave the argument about differences in economic environment and the manufacturing capacity of the technology-receiving country for later. A key point for successful technology transfer is the approach taken to enforce the system and process design to produce good products. The transfer of conditions alone will not create a good product. These conditions differ once the product, equipment size, or quantity of production are varied. The condition does not create the product, nor is the product produced to follow the condition. The condition of the equipment is determined so as to obtain a good product. I would like to suggest that Chinese engineers examine and change their approach (attitude or philosophy) toward engineering.

S. Does this mean that the Chinese cannot produce good products until they change their approach, attitude, and philosophy about engineering?

E. Yes it does. China is attempting to accumulate technological capability through the introduction of the technology of advanced countries, technical cooperation arrangements, and by sending engineering students abroad.

However, what determines technological capability and manufacturing capability is the approach taken by the people who are engaged in the technology and manufacturing (engineers as well as managers and directors).

China will remain a third-class country as long as it maintains its present attitude.

S. I am impressed with the deep insight you obtained from only two trips to China, but it does seem to be an oversimplification based on a small number of observations.

E. That may be so. But let us go back to the original question about science and engineering. The Chinese put the blame for inferior products on 1) the material and 2) equipment.

S. I think it is natural for the Chinese to have these characteristics because they are hard-working. I think any scientist shows interest in such things. Isn't it natural to ask about experimental conditions in detail?

E. I am not saying whether it is good or bad. I simply wanted to explain why the Chinese cannot produce good products.

S. I understand. Please go on.

E. Simply speaking, Chinese engineers are "scientists and not engineers."

S. Are scientists and engineers different? Please give me your definitions.

E. **Scientists analyze a phenomenon and try to determine its regularity.** Usually such people are not found in factories or companies.

S. I understand that. But what is the definition of an engineer? I thought an engineer is a scientist working in a factory. Am I wrong?

E. That is the concept of an engineer generally accepted by the Japanese. But this is a false perception. **An engineer is "a person who designs the process of producing a product that performs the objective function at the minimum cost."**

S. A person who designs the process? I cannot accept that. Isn't he a person who creates a good product?

E. A good product is produced in a good process. That is why we have to consider the process, not the material. This is an important point.

S. I am afraid I do not understand what you are saying. Can you put it a different way?

E. I know you are a scientist. What do you do if you want to produce a good product?

S. I think a good product is produced by a skilled worker with good material and a good machine.

E. I am afraid that China does not have good materials or good machines.

S. Bad materials, bad machines, and unskilled workers only produce inferior products.

E. Actually, they are producing inferior products with bad machines composed of bad parts.

S. It sounds like a desperate, vicious cycle.

E. There are many countries that have experienced similar paths.

S. Japan right after the war was in a worse situation, but Japan is producing good products these days. How did they turn it around?

E. You bring up a good point. The recovery of the Japanese manufacturing industry was rather rapid.

Although the economy suffered from a shortage of materials, the time when inferior products could sell did not last long. We believe that this is due to the fact that only good products can survive in a free market in a competitive society, and this encourages production with good materials and good equipment. Inferior products were eliminated in the process of natural selection. Though I can think of other reasons, the most important factor in Japan compared to China was the existence of free competition.

S. Only good products can survive through a competitive free market.

E. Yes, indeed.

S. Since there is no competition in China, it is difficult for the Chinese to create good products, is that it? How can a Chinese product become competitive in a free market?

E. Let us limit our discussion only to the technological aspect by setting aside the problem of the free market's introduction for the time being.

What the Chinese need to do is to produce good products from bad material, with bad equipment, and unskilled workers.

S. That is scientifically impossible. What we can create from bad things are bad things. That is self-evident. Your statement is rash.

E. **It may be correct scientifically, but wrong from the engineering viewpoint.**

S. Do you have any examples at hand?

E. I have an example of a plating process at the time of new factory start-up. Uneven chrome plating was found on plated parts. It seemed that this was caused by a composition change in the plating solution over time, as well as moisture and temperature changes.

S. We often see that. In such a case, we should improve the instrument control conditioner. Also, the on/off switch frequency for increased heat exchange capacity must be made to keep the bath temperature constant. What about that?

E. I think we should *not* do such things.

S. Why not?

E. I think that we should not act upon and attempt to control things that have variability. **The only thing we need to do is reduce the impact of variability on the product. It costs a lot of money to control variability.**

What you are suggesting is the last and worst improvement method we can take. We should use it only when there is no better solution.

S. What did you do then?

E. We extended the distance between the electrodes by 1.5 times and doubled the agitation flow quantity, together with adding bubble agitation. The objective was achieved by these methods. It cost only ¥50,000, about one hundredth the cost of what you have suggested.

S. I think you were simply lucky to eliminate the problem with such simple methods. Wasn't your success just due to chance?

E. Only God knows whether I was lucky or not, and I have nothing to do with God. I simply had a method for finding these solutions quickly, securely, and economically.

S. What is the name of this method?

E. It is called the experimental design method or Quality Engineering. This method enables us to find luck even if we are not normally lucky.

S. It sounds interesting. Can you teach me this method?

E. Of course I can, but I am too busy today, so make an appointment for another time. All factories have traditional examples of this kind, the solution to which can cost a lot of money. This method should be mastered by corporate engineers if they want to become successful. I believe that this method was born out of the accumulated wisdom of many engineers. It is often referred to as Taguchi Methods in North America and Europe. The method is, of course, backed by mathematics.

S. I understand. But about the example of plating, did you find the cause of the uneven plating? Can you tell me how far the analysis went?

E. **We did not find the cause, and we did not even attempt to find it.** Of course, we did not do any analysis. Analysis often increases the volume of unnecessary information, resulting in further chaos. Although I do not pursue cause, I recommend establishing these hypotheses. Hypotheses often bring in more useful information than knowledge about the true cause.

S. I cannot accept that. It does not make sense to take a countermeasure when the cause is not known.

E. Oh, is that so?

S. Yes. First of all, such a method lacks scientific backing. It is not supported by scientific facts, and it may not even coincide with the law of nature. It is a fraud.

E. Is that really so? Look at the results. A stable product with high quality was supplied to the market in a short period of time. It required only one one-hundredth of the potential cost. The objective was achieved.

S. I know that the objective was achieved, but I cannot accept this idea easily. I think it is not scientific.

E. **What does it mean to be scientific?**

S. In this case it means **reasoning about the phenomenon according to the laws of nature.**

E. I do not think it is necessary to reason about the phenomenon.

S. Without reasoning, we cannot correctly interpret and explain the phenomenon. I believe that this is the basis of scientific attitude.

E. It may be natural to think that way in the scientific field, but what we are discussing is the engineering field. Engineering is supposed to recommend the most effective way to improve our productive capacity. I know many engineers who think like you, but these people are not successful in their corporate careers and are causing confusion and loss to their companies.

S. Why is that?

E. **It is because the phenomenon exists before reasoning and laws.** An apple fell to the ground before Newton discovered why. The best mathematical model is one that coincides with the phenomenon. Some people often conclude that a fact is false if it does not follow the law and the model. **I think that the majority of phenomena in the world cannot be explained theoretically. The fact that a large number of so-called scientists still exist is the best proof of that.**

S. Well, I agree that the majority of phenomena remains unexplained. How, then, do the scientific attitude and the engineering attitude interact?

E. When we want to utilize a phenomenon, we need an engineering attitude. On the other hand, when we research regularity to create a model, we need a scientific attitude.

S. Now I can see that it is possible to produce products with uniform quality even if the equipment is bad and causes large variability.

E. The same thing can be said in the case of bad materials.

S. Do you have any other examples?

E. Well, I can give you an example in which a scientist and an engineer were compared. It happened in an electronic-parts factory.

A new gloss-plating line was started in a factory, but the product was defective, with imperfectly glossed spots. Even the research laboratory became involved in the analysis of the gloss and non-gloss spots. Analysis results indicated that a difference existed in the crystal structure of the spots. It was found, through element analysis, that the difference in the crystal structure was caused by the difference in the plated layer.

Furthermore, the actual number of cases that generated the difference in metal distribution of the plated layer were counted. It was decided that a more detailed analysis of the previous process and the raw material should be carried out. It took another month to do this, and then the research lab realized that an additional month would be necessary to analyze the results. On top of all this, the person in charge of the analysis said that he might or might not discover the cause in his analysis, and that finding the cause and taking countermeasures were two different things.

The sales department became irritated with this very slow process. When everyone had become thoroughly discouraged, one engineer proposed the following: "**The cause could be in the previous process or in the raw material. But regardless of the surface condition, what we need to do is produce a uniform quality product without spot defects.** Let us try to achieve uniform quality by changing the plating condition."

Because the line was already idle, everyone agreed to try this out. As a result, they came up with using increased electrical current, increased temperature, and reduced density. This made it possible to ship out the final product. This engineer was not even a plating expert.

S. You have introduced two different types of technology experts. The former is a scientist and the latter is an engineer, is that right?

E. Yes, that is true.

S. What is the difference?

E. A scientist is inclined toward the material, whereas an engineer is inclined toward the process.

S. What is the difference between material inclination and process inclination?

E. A technology expert inclined toward the material tries to maintain a process condition that generates a good product. But this will produce slightly different products every time the process changes even slightly. This leads to a lack of process stability, and good products cannot be produced.

S. Is the scientist wrong in doing this?

E. Yes, he is. In this example, he committed at least three errors.

The first was the choice of method for deciding the optimum process condition. The scientist determined the optimum electrolyte condition based on the precision test described in *Table 1-6*.

Table 1-6: Electrolyte Condition

Levels \ Factors	Temperature (°C)	Density (%)	A/dm²	Flow Speed (m/min.)
1	T_1 (40)	C_1 (3)	A_1 (5)	Q_1 (1)
2	T_2 (50)	C_2 (4)	A_2 (8)	Q_2 (2)
3	T_3 (60)	C_3 (5)	A_3 (11)	Q_3 (3)

The optimum condition of T was determined to be T_3 by fixing the levels at C_2, A_2, and Q_2. That of C was determined to be C_1 by fixing the levels at T_3, A_2, and Q_2. That of A was determined to be A_2 by fixing the levels at T_3, C_1, and Q_2, and that of Q was determined to be Q_3 by fixing the levels at T_3, C_1 and A_2. This gives an optimum condition of T_3, C_1, A_2, and Q_3.

S. That is a regularly used method. We do the same thing.

E. Yes, but this method is risky. When the order of level determination is different from *TCA* and *Q*, the result may be different. Actually, another result indicated the combination of T_3, C_1, A_2, and Q_3 to be the optimum condition.

S. Yes, that can happen. But we perform experiments in universities and write articles using that method.

E. Isn't it strange that the optimum condition is different depending on the order of the experiments?

The purposes of experimentation in corporations and in universities are quite different. The activity of a scientist is to analyze a phenomenon and discover and publish its laws. Therefore, a scientist uses a precision test to discover regularity and to estimate cause from the changes that took place in the causal system due to a given impact.

This is why the precision test is acceptable in university research.

S. Well, I understand that very well. Could you explain the purpose of tests in corporations?

E. The following three test steps are used in corporations:

1. System design

   This determines which function and mechanism should be utilized.

2. Parameter design

   This is an experiment to achieve the target quality by determining the conditions that allow for target quality. Here the system condition (levels) to fulfill the objective function is determined.

   This includes the experiment that allows us to determine the condition for achieving the same function, even when the conditions of raw material, production environment, and usage environment are changed. In the above example, such an experiment was not included. The process should have been designed so as to achieve the same function, regardless of the variability caused by the previous process or by the raw material.

3. Tolerance design

   This gives the rational grade limit for components of a system (i.e., the elements, parts, and conditions). For example, this includes the electronic parts grade, the level of finishing of a mechanical part, and the control limit of a process condition.

S. Experiments done by scientists are intended to discover regularity by eliminating variability, so as to single out the effect of one factor. This is why other factors are held constant.

E. I cannot tell which test is best for discovering regularity. I think you scientists know more about this. What I want to emphasize here is the fact that **the ordinary precision test does *not* provide us with a process that satisfies the target quality function and whose quality is resilient against changes in the environment.**

S. That sounds quite right, but I was not taught such a method at the university. What is the method of process design that provides products with strong resilience?

E. I have already mentioned the name, i.e., the experimental-design method of Quality Engineering. The Signal-to-Noise ratio is a metric used in this method that is particularly useful for process design. This method of experimental design is widely used in Japan and is also increasingly being used in the United States, China, and India.

S. I think I now understand the purpose of the experimental-design method.

E. Oh, do you?

S. Can you go back to your story about Chinese engineers?

E. Oh, I am sorry, I lost track.

   It is true that their raw material and equipment are inferior. Under such conditions, good products cannot be produced if they adhere to scientists' ideas. **I want to insist that engineers should understand what the role of an engineer is and that they be inclined toward process design.**

It is true that all of China is enthusiastic about absorbing technology. But the transfer of scientific and technological achievements is overemphasized. They should also absorb approaches, attitudes, and a philosophy toward process design technology, production, and inspection.

S. Now I agree with you.

2

Traditional Experimental Design Deficiencies

2-1. Four Methods for Determining the Optimum Condition and Their Limitations

Q. In new product development, equipment design, and process design, the optimum combination of factors and levels must be chosen from among many factors and levels. An engineer spends a great deal of time and energy in making such a selection. What method should he use?

M. *Table 2-1* compares four methods for determining the optimum condition for a case of four variables with three levels.

Table 2-1: Comparison of Four Methods

	Name of Method	Content of Method	Evaluation	
			No. of Experiments*	**Reliability**
1.	Multi-factor layout	Every combination of levels tested	Many (81)	High
2.	One-factor-at-a-time experiment (Precision test)	Effect of each factor confirmed (one-at-a-time) by fixing levels of other factors	Several (12)	Low
3.	Target method	Only 2 or 3 possible "best" combinations of levels (2 or 3) tested	Few (2–3)	Lowest
4.	Experimental design method	Experiment using orthogonal arrays	Several (9)	Highest

*The number of experiments needed to determine the optimum condition for four factors with three levels (e.g., 81).

These four methods are often used by the engineers. The one-factor-at-a-time method is the most frequently used.

The **multi-factor layout** is also called the **all combination method** or **the full-factorial method**. This method is thought to have the highest probability of locating the optimum condition. However, this method requires many experiments, and is known to cause a lot of problems for engineers.

Q. The multi-factor layout method gives highly reliable results, but is not practical because of the large number of required experiments. This usually causes engineers to select another method, one that requires only a limited number of experiments. Examples of such methods are the one-factor-at-a-time experiment and the target method.

M. The **one-factor-at-a-time method** (also called the **precision test method)** is widely used for optimization of many factors and levels for a new product. It is a method for obtaining the impact of a factor by changing the level of that factor, while maintaining the levels of other factors at their most reliable levels.

On the other hand, the **target method** is used to confirm whether the target level is contained in the optimum combination levels proposed by the engineer. An important characteristic of this method is that it will require the smallest number of experiments when everything goes well.

What engineers really want is to reduce the number of experiments without losing the reliability of experimental results. An engineer will accept any method that provides highly reliable results from a small number of experiments.

Q. So, the problem is how to reduce the number of experiments while maintaining the reliability of experimental results, isn't it? But I have never heard of a method that can do this.

M. Most engineers are troubled by the fact that they cannot obtain highly reliable information from their experiments in order to correctly judge whether to "stop or go."

Q. I can understand that point, but what is the problem with the one-factor-at-a-time method?

M. One engineer determined the optimum condition of the four factors with three levels shown in Table 1-6 using the one-factor-at-a-time method. What he got was T_3, C_1, A_2, and Q_3. This was based on the assumption that the factors on the left side have a larger effect, and by fixing the levels of the right-side factors at two when he changed the level of the left-side factors.

Q. Can't we rely on the combination T_3, C_1, A_2, and Q_3 obtained that way?

M. This optimum condition was determined by using the factors in the order from the left side to the right side shown in *Table 1-6*. But what happens if we start with the factors from the right side and go on to the left side?

Q. According to my experience as an engineer, it is rather rare to obtain the same combination of levels as the optimum condition. This is because of the interaction among factors.

M That is quite right. The optimum condition obtained in the reverse order was T_1, C_3, A_1, and Q_1. By definition, there should be only be one optimum condition for a system. **It is curious that the optimum condition changes when the order of the experiments is changed.**

The engineers themselves, the persons who use the one-factor-at-a-time experiment for optimum condition determination, know quite well how unreliable a condition obtained from this experimental method is, and how large a loss they can cause the company because of the lack of reproducibility of results. This occurs because many defective products, with a characteristic value far different from what was expected, are often created when the optimum condition determined by a one-factor-at-a-time experiment is applied to a production machine. This requires additional repeated experiments.

It is the engineers who determine the optimum condition using the one-factor-at-a-time method who experience this kind of thing. The thing that can save engineers from this vicious circle is an experimental design method supported by new concepts and a new philosophy.

2-2. Education in Experimental Methodology for Engineers

Q. Why then do engineers use the one-factor-at-a-time method to determine the optimum condition? Is there any other situation in which the one-factor-at-a-time method is used?

M. **Engineers use the one-factor-at-a-time method because they know of no other method for optimum condition determination for many variables with many levels.** Engineers, of course, receive a higher education. In the final year of college, all engineers must write a graduation thesis. The graduation thesis gives engineering students their first opportunity to perform an experiment based on their own ideas.

The instructor teaches a student to first find the general tendency when one variable is changed while the other factors are kept constant (the relationship between the characteristic value and the changes in the level of that factor) and then to confirm how the characteristic value changes by combining the most favorable levels. This is what college instructors teach students.

Q. So, engineers learn the one-factor-at-a-time method in college, don't they?

M. That is right. **All of the experiments used in research reported in academic journals are done by the one-factor-at-a-time method.** This is because the purpose of academic research is to compare the researcher's discovery against conventional knowledge and to stress the difference under the most simplified condition. What is necessary there is to perform experiments that are comfortable to the researcher and that make it easier for others to understand. This is why universities only do one-factor-at-a-time experiments, and it is also the reason that this method is widely accepted as the general experimental method.

Because activities in universities are not economically competitive activities, the selection of levels and the determination of the optimum condition do not cause any direct loss or problems with the research. The absence of loss or problems does not encourage changes or improvements in the experimental method. All researchers know that it is very difficult to reproduce the absolute value or experimental data reported in an article.

Q. This means that **the one-factor-at-a-time method itself is essentially unsuitable as an experimental method for obtaining data reproducibility, process stability, and reliability**, doesn't it?

M. Yes. Since it is the experimental method that is cultivated in university education and university research, it cannot respond to the stability and reliability requirements of industrial production. However, even though this method provides data that do not pass stability and reliability requirements, we should not fault universities for teaching this method. The university is a place for educating scientists, and it has nothing to do with the world in which engineers are exposed to the constraints of corporate cost-benefit relationships.

Q. Then **it is the corporations that must educate engineers**. Is that right?

M. Yes, indeed. Every corporation is responsible for educating its engineers in an experimental method that provides designs with low cost and high reliability in a short period of time. They must do this in order to survive corporate competition. Engineers have a high potential, and corporations must guide them so that their potential is maximally utilized. It is unreasonable for a corporation to complain about the ability of an engineer without providing this kind of training.

Q. But I am afraid only a few companies educate engineers on the basis of the new philosophy.

M. That may be true. And this is the time for them to revise their engineering training.

2-3. A Method for Determining the Optimum Condition

The optimum condition in *Table 1-6* was different depending on whether the order of the factors was taken from the right or left. What method should be used to determine the optimum condition?

Q. Suppose we create a new system composed of two level factors *A*, *B*, *C*, and other factors. Suppose further that the first and second levels of each factor are tested under various conditions. The average data of the first and the second levels, as well as the differences, are summarized in *Table 2-2*. Here, the larger data values indicate a better result. According to these data, how can we evaluate the size of the effect of factors *A*, *B*, and *C*?

Table 2-2: Average Data Difference between (A_1, A_2), (B_1, B_2), and (C_1, C_2) (From Experiments with Various Level Combinations)

	(1)	(2)	(3)	(4)	(5)	(6)	(Average)
A_1	73	60	58	65	47	58	60.2
A_2	63	42	51	55	39	43	48.8
Difference	+10	+18	+7	+10	+8	+15	11.4

	(7)	(8)	(9)	(10)	(11)	(12)	(Average)
B_1	59	53	60	50	52	52	54.3
B_2	50	55	54	49	47	56	51.8
Difference	+9	–2	+6	+1	+5	–4	2.5

	(13)	(14)	(15)	(16)	(17)	(18)	(Average)
C_1	48	53	55	53	56	54	53.2
C_2	51	51	54	57	53	56	53.7
Difference	–3	+2	+1	–4	+3	–2	–0.5

( ) = Combination number of various levels

M. I think it will be in the order *A*, *B*, and *C*. This is because the larger difference between levels indicates the larger factorial effect. The level A_1 of factor *A* is good in every experiment. Though the level B_1 is not good in every experiment, we can judge that B_1 is better from the result of many experiments.

For *C*, neither level appears to be better. In this case, we can say that neither level will cause a particular loss.

Q. The difference in factorial effects is different, depending on the combination of levels A_1, A_2, B_1, B_2, C_1, and C_2. Why don't we get the same value?

M. The difference between A_1 and A_2 is not necessarily the same under different conditions. In addition, the existence of interactions, large or small, gives different levels of difference for each experiment. **Because every experiment gave a difference between A_1 and A_2 in the laboratory, we can expect a main effect that reproduces this difference, even in mass production (by an actual production machine).**

For *B* and *C*, the difference between levels may not appear in mass production because their main effects are larger than their interaction by only a small margin. Their effects may be as large as the interaction. In other words, only those levels of **main effects** that are far larger than the interactions can be reliably realized in mass production.

Q. Does that mean that interactions are part of the so-called noise component?

M. Yes. We should consider **the interaction itself to be the noise component. In other words, experimental results can be trusted only when interactions are small.** When interactions among factors are large, the reliability of the experiment drops markedly.

In the case of an experiment in which the interaction is expected to be large, the following methods should be used.

1. Selection of factor levels that minimize expected interaction among factors
2. Appropriate data transformation.

M. Then we can determine the optimum condition for many factors with many levels using the following steps, can't we?

Step 1 Perform experiments at the level one wants to compare under various combinations of levels.

Step 2 Select the levels of the factors that have a strong influence on the characteristic value based on a comparison of the characteristic values they yield. The levels of factors that have a medium-to-small influence on the characteristic value are determined by considering other objective characteristic values, cost, and productivity.

Q. If these steps are taken, level A_1 turns out to be better for factor A. The best levels of B and C can be determined by considering the other characteristic values. In order to achieve a better major characteristic value, B_1 is better for B, but the level of C does not matter here. The level of C should be determined by checking which level gives a lower cost and better operating conditions. The graph shown in Figure 2-1, which shows the average values of A, B, and C, is called the **factorial effects chart**.

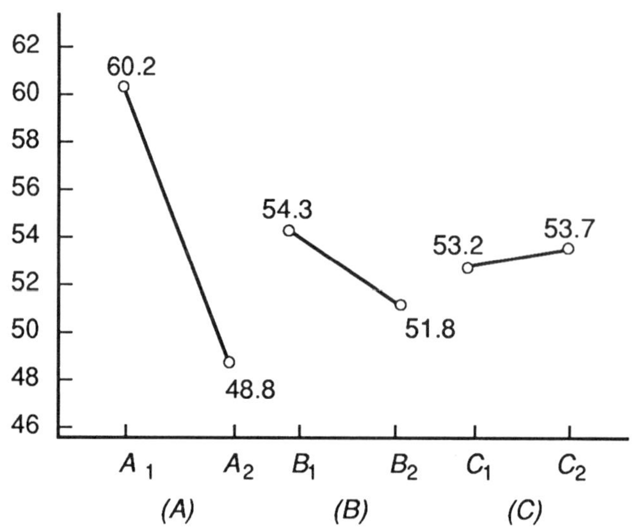

Figure 2-1. Factorial Effects of A, B, and C

M. That is right. The objective characteristic value must be used primarily to choose the level of factors whose effect is large. Even if the effect on the objective characteristic value is medium or negligibly small, the level that gives the better value should be selected. However, because the impact on the characteristic value is small for factors with a medium-to-small effect, other levels may be chosen based on other reasons, such as cost and productivity.

2-3-1. Experimentation Using Orthogonal Arrays

Q. The method for determining the optimum condition for many variables with many levels is shown in *Table* 2-2. It is based on experiments run under many different combinations of levels and may give reliable effects. But this method is not practical because it requires too many experiments.

M. *Table* 2-2 is used to check factors *A*, *B*, and *C* under various conditions in order to verify that the levels selected on the basis of the experimental results show good reproducibility. In practice, the optimum levels must be determined using a practical and limited number of experiments. To accomplish this, a special table called an **orthogonal array** is used.

Q. What is an orthogonal array?

M. An orthogonal array experiment must satisfy the following conditions:

1. The experiment must be conducted under different combinations of factor levels.
2. The number of experiments required must remain within practical limits.
3. Comparison of factor levels must be balanced.

Let us consider an experiment using three factors (*A*, *B*, and *D*) with two levels with the multi-factor layout. *Table* 2-3 shows the factor and levels, and *Table* 2-4 shows the actual data for the multi-factor layout experiment. As *Table* 2-4 indicates, there are eight combinations. *Table* 2-5 is an experiment instructions form corresponding to the factors and levels. Table 2-6 shows the factor number and level name of experiments corresponding to each entry in *Table* 2-5.

Table 2-3: Factors and Levels

Factor	Factor Symbol	Level #1	Level #2
chromic acid anhydride (g/l)	A	400 (A_1)	350 (A_2)
sodium hydroxide (g/l)	B	58 (B_1)	50 (B_2)
sulfuric acid (g/l)	D	0.75 (D_1)	1.50 (D_2)

Table 2-4: Multi-Factor Layout for *A, B, D*.

		Sulfuric Acid (g/l) *D*	
Chromic Acid Anhydride (g/l) *A*	**Sodium Hydroxide (g/l) *B***	**0.75 (D_1)**	**1.50 (D_2)**
A_1 400	B_1 58	No. 1 y_1	No. 2 y_2
	B_2 50	No. 3 y_3	No. 4 y_4
A_2 350	B_1 58	No. 5 y_5	No. 6 y_6
	B_2 50	No. 7 y_7	No. 8 y_8

Table 2-5: Experiment Instructions

Factor / No.	Chromic Acid Anhydride (g/l) *A*	Sodium Hydroxide (g/l) *B*	Sulfuric Acid (g/l) *D*	Data
1	400	58	0.75	y_1
2	400	58	1.50	y_2
3	400	50	0.75	y_3
4	400	50	1.50	y_4
5	350	58	0.75	y_5
6	350	58	1.50	y_6
7	350	50	0.75	y_7
8	350	50	1.50	y_8

Table 2-6: Experiment Instructions Based On Levels

Factor / No.	*A*	*B*	*D*	Data
1	1	1	1	y_1
2	1	1	2	y_2
3	1	2	1	y_3
4	1	2	2	y_4
5	2	1	1	y_5
6	2	1	2	y_6
7	2	2	1	y_7
8	2	2	2	y_8

Q *Table 2-6* appears to be a convenient table because it shows the levels for each experiment number.

M There is another table that shows corresponding levels for each experiment and serves a purpose similar to that of *Table 2-6*.

This table or matrix is **the orthogonal array**. Examples of such arrays are given at the end of this book. Table 2-7 shows an **L_8 orthogonal array,** which has 8 combinations of levels. Each column in this array corresponds to a factor. An experiment with seven factors, with two levels each, can be run using an L_8 (2^7). Let us construct an experiment using an L_8.

Table 2-7: L_8 Orthogonal Array

Factor / No.	1	2	3	4	5	6	7
1	1	1	1	1	1	1	1
2	1	1	1	2	2	2	2
3	1	2	2	1	1	2	2
4	1	2	2	2	2	1	1
5	2	1	2	1	2	1	2
6	2	1	2	2	1	2	1
7	2	2	1	1	2	2	1
8	2	2	1	2	1	1	2

Let us devise a research project for the improvement of chrome plating in which the optimum condition for seven factors at two levels must be determined. Optimization will be based on such characteristics as wear rate, plated film thickness, adhesiveness, glossiness, thickness uniformity in plating, and appearance. *Table 2-8* summarizes the factors and levels. First, make a correspondence between the seven selected factors, *A* – *G*, and columns 1 – 7. Next, make a correspondence between the numbers 1 and 2 in each column and the actual levels of each factor to be run in the experiment, so as to determine the eight combinations numbered 1 – 8. This process is called **assignment.** Our chrome plating experiment will be done for these eight combinations of levels. *Table 2-9* shows the assignment and the results of this experiment. Fifty samples are plated simultaneously for each experiment, and the number of samples indicated for each characteristic is taken and evaluated.

Table 2-8: Factors and Levels — Chrome-Plated Film Experiment

Factor Symbol	Factor	Level	
		1	2
A	Chromic acid anhydride (g/l)	400	350
B	Sodium hydroxide (g/l)	58	50
C	Trivalent chromium (g/l)	7.5	10.0
D	Sulfuric acid (g/l)	0.75	1.50
E	Bath temperature (°C)	18	23
F	Current density (A/dm^2)	30	45
G	Liquid circulation flow rate (l/min.)	20	40

Table 2-9a: Assignment and Experiment Results of Chrome-Plated Film

No.	*A* 1	*B* 2	*C* 3	*D* 4	*E* 5	*F* 6	*G* 7	Chromic Acid Anhydride (g/l) 1	Sodium Hydroxide (g/l) 2	Trivalent Chromium (g/l) 3	Sulfuric Acid (g/l) 4	Bath Temp. (°C) 5	Current Density (A/dm²) 6	Liquid Circulation (l/min.) 7
1	1	1	1	1	1	1	1	400	58	7.5	0.75	18	30	20
2	1	1	1	2	2	2	2	400	58	7.5	1.50	23	45	40
3	1	2	2	1	1	2	2	400	50	10.0	0.75	18	45	40
4	1	2	2	2	2	1	1	400	50	10.0	1.50	23	30	20
5	2	1	2	1	2	1	2	350	58	10.0	0.75	23	30	40
6	2	1	2	2	1	2	1	350	58	10.0	1.50	18	45	20
7	2	2	1	1	2	2	1	350	50	7.5	0.75	23	45	20
8	2	2	1	2	1	1	2	350	50	7.5	1.50	18	30	40

Table 2-9b: Assignment and Experiment Results of Chrome-Plated Film — Raw Data, Type of Characteristic, and Appropriate Data Transformation

No.	Wear Rate (mg)			Plated-Film Thickness (μ)					Adhesive-ness (g)		Glossi-ness	Thickness Uniformity in Plating			Appear-ance Judgment		
												* G	M	B	G	B	
1	8	7	9	9.3	8.3	8.9	11.7	12.5	10	15	52	4	1	0	4	1	
2	9	7	10	10.6	9.5	10.3	10.0	11.8	12	9	51	1	3	1	5	0	
3	6	5	3	8.5	8.1	11.2	12.5	12.3	6	3	25	1	0	4	5	0	
4	2	3	3	12.5	7.7	10.2	12.3	7.6	9	12	34	0	3	2	2	3	
5	3	2	1	10.5	7.9	11.3	12.3	8.7	5	7	69	0	0	5	2	3	
6	31	28	35	7.6	6.0	12.5	12.1	13.0	4	6	77	1	2	2	0	5	
7	21	18	25	7.1	7.7	7.9	13.9	13.1	8	6	86	2	1	2	2	3	
8	13	21	10	9.8	10.3	8.6	12.6	7.6	10	14	95	0	0	5	0	5	
	Smaller-the-better			Nominal-the-best					Larger-the-better		Simple continuous variable	Classified continuous variable			0/1 data (discrete data)		Type of characteristic
	S/N ratio			S/N ratio					S/N ratio			Accumula-tion			Accumula-tion		Data transforma-tion

* G = Good
  M = Medium
  B = Bad

Table 2-5 is the same as an L_8 orthogonal array, with *A, B,* and *D* corresponding to columns 1, 2, and 4.

Q. Could you first explain the terms used in referring to an orthogonal array?

M. An orthogonal array can be discussed as follows (See *Table 2-9a*): An L_8 (2^7) array is simply referred to as an L_8. The number designation (No.) in each horizontal row is called the **experiment number** or the **assignment number.** On the other hand, vertical columns are called the **columns** of the orthogonal array and correspond to the factors. The numbers 1 and 2 in each column indicate the levels of the factors. Each column has an equal number of 1s and 2s. Any two columns are said to be balanced if the combinations. (1, 1), (1, 2), (2, 1), and (2, 2) appear the same number of times in each column. These columns are then said to be **orthogonal.**

Q. Does an experiment using an orthogonal array satisfy the three conditions mentioned earlier?

M. The orthogonal array satisfies those conditions. In the example for *Table 2-9*, the combination of levels to the right of A_1 and A_2 is composed of four different combination of *B, C, D, E, F,* and *G*. **Because B_1, B_2, C_1, . . ., G_1, G_2 appear the same number of times for A_1 and A_2, *A* is orthogonal or balanced, and we can choose either level, A_1 or A_2, by comparing their average values.** Because A_1 and A_2 are examined under different conditions, we can choose a reliable level that will be valid under different conditions.

Q. This means that, if we look at experiments that have the same level of a particular factor, we can find the equal number of levels of other factors in the orthogonal array, doesn't it?

M. Yes, it does. As I mentioned earlier, we call such a table orthogonal when the levels of each factor correspond with good balance.

Q. How effective is the experiment shown in *Table 2-9* when compared with the multi-factor layout?

M. If we use an L_8 (2^7), and if there are no repeated experiments, eight samples will provide the same information as that obtained from the (2^7 or 128) samples needed for the multi-factor layout.

There are many kinds of orthogonal arrays. *Table 2-10* lists some arrays and the number of experiments required when a corresponding multi-factor layout is used.

Table 2-10: Comparison of Orthogonal Arrays: Number of Factors Assigned and Number of Experiments

Levels	Orthogonal Array	No. of Factors Assigned		No. of Experiments	No. of Experiments for Same Info. with Multi-Factor Layout
		2-level	3-level		
2-level	L_8	7		8	128
	L_{12}	11		12	2,048
	L_{16}	15		16	32,768
	L_{32}	31		32	2,147,483,648
3-level	L_9		4	9	81
	L_{18}	1	7	18	4,374
	L_{27}		13	27	1,594,323
	L_{36}		13	36	1,594, 323

Even the largest orthogonal array, L_{36}, in *Table 2-10,* requires only 36 experiments. This means that the number of experiments is within practical limits. An experiment using an orthogonal array is thus both attractive and efficient. It gives us reliable results that have been examined under different conditions using a feasible number of experiments in terms of cost and time.

2-3-2. Different Orthogonal Arrays and Their Use

Q. It seems as if there are many types of orthogonal arrays. Could you explain their differences and their uses?

M. Orthogonal arrays can be classified on the basis of the number of assigned factors and levels, as well as to whether they check interactions. They are summarized in *Table 2-11.*

Table 2-11: Orthogonal Array Classification

Type of System	Ways in Which Interactions Can Be Treated	
	Interactions can be obtained from columns corresponding to lines on linear graphs.*	Effects of interactions cannot be obtained because interactions are confounded in each column. Columns give information on main effects only.
2-level series	L_4, L_8, L_{16}, L_{32}	L_{12}
3-level series	L_9, L_{27}	L_{18}, L_{36} **

* For orthogonal arrays that can have multi-levels.
** These arrays belong to mixed series because some columns have two levels.

Q. How do we select an appropriate orthogonal array?

M. First, selection of the two- or three-level system is made based on the number of the factor levels one wants to test. Four and eight levels are considered to belong to the two-level system. Nine levels are considered to belong to the three-level system. A four-level system can be treated as a three-level system and a three-level system can be treated as a two-level system, using a technique called the dummy method. They are then considered to belong to three- and two-level systems, respectively.

When we want to experiment with interactions or a number of levels larger than the normal levels, we use the arrays listed on the left side, i.e., L_4, L_8, L_{16}, L_{32}, L_9, and L_{27}. When we want to obtain only main effects, which are not influenced by interactions, we use those arrays listed on the right side of *Table 2-11*, i.e., L_{12}, L_{18}, and L_{36}. Finally, selection of the actual array to be used is based on the necessary quantity of information in the factors, levels, and interactions the engineer chooses to study (i.e., the number of main effects and interactions).

Q. It sounds complicated.

M. I recommend using L_{12}, L_{18}, and L_{36} arrays, even for beginners. Although the number of assigned levels is fixed, they are easy to use and we can obtain reliable information. The L_{18} and L_{36} are particularly convenient because we can assign factors with two levels to the three-level system using the dummy method.

2-3-3. Precision Experiments and Orthogonal Array Experiments

Q. In order to obtain seven-factor information similar to that obtained from an L_8, we can utilize the type of experiment shown in *Table 2-12*—a typical precision (one-factor-at-a-time) test method. The main effect of *A* can be obtained by comparing No. 1 and No. 2, and the main effect of *B* can be obtained by comparing No. 3 and No. 4, . . ., etc.

Table 2-12: Precision Experiment

No.	*A*	*B*	*C*	*D*	*E*	*F*	*G*
1	1	1	1	1	1	1	1
2	2	1	1	1	1	1	1
3	2	2	1	1	1	1	1
4	2	2	2	1	1	1	1
5	2	2	2	2	1	1	1
6	2	2	2	2	2	1	1
7	2	2	2	2	2	2	1
8	2	2	2	2	2	2	2

M. This method is not good at all because the difference between levels is tested only once, and the backgrounds are not equal. In actual production, where the background changes greatly, this method gives output that is completely off the target value, with large variability. A_1 and A_2 in orthogonal table L_8 are measured from the result of four experiments conducted under different conditions. The results will be reliable.

Q. But the effect of A_1 and A_2 in the case of *Table 2-12* can be obtained precisely without error from No. 1 and No. 2 because factor *A* was the only item changed from experiment No. 1 to No. 2. I have been told that data without error are good data. The L_8 experiment in *Table 2-9* involves many level changes from one experiment to the next, generating much error. It cannot be called precise.

M. Between-experiment error is smaller for *Table 2-12* if we consider the precision of each experiment. **But the engineer is not looking for an experimental method that reduces experimental error. His objective is to select a combination of levels that will be robust against the changes in the environment when the optimum condition is applied.**

Therefore, the optimum condition obtained from *Table 2-9*, where the experimental background varies greatly is more reliable than the optimum condition obtained from *Table 2-12* where the experimental condition is precisely determined. In other words, the data obtained from *Table 2-9* can be reliably applied as a good optimum condition on an actual production floor.

2-4. Traditional Experimentation—Fisher's Three Basic Principles

Q. **The premise of traditional experimentation is that an experiment must be conducted according to three basic rules, i.e, repetition, randomization, and local control, before significance level testing.** These three fundamental principles and significance level testing are the basis of traditional experimental planning, aren't they?

M. It is true that Fisher's Three Basic Principles and test have been the basis for the development of experimental design in Japan. But now is the time to reconsider the constraints imposed by the Three Basic Principles and test.

Fisher applied the experimental method mainly in the development of agricultural and medical experiments. But what we are concerned with are industrial experiments. These are different from agricultural experiments, which compare plants in the natural environment. Let us examine the application of Fisher's Three Basic Principles in the context of industrial experimentation by comparing agricultural and industrial experiments.

An agricultural experiment is characterized by the fact that it faces a large noise component called nature. It has a small number of controllable factors. Because agriculture affects the entire globe, even a small effect is important to the human race.

For example, if preprocessing increases the germination rate of one kind of seed by 1%, we can expect that the entire output of that plant will increase by 1%. Furthermore, if this seed is effective only under a certain condition, say it can grow only in temperate zones, we can increase the output by using the preprocessed seed in a temperate zone.

This means that there is an interaction between preprocessing and the cultivation area. When we perform this kind of experiment, we expect to obtain not only the main effect, but also the interaction by separating out the noise component. **In an agricultural experiment, we need an instrument to judge whether the main effect and the interaction are effective against the noise component.**

Q. That instrument is significance testing, isn't it?

M. Yes, it is. The same thing applies to medical experiments. If a drug is effective for certain patients, it can save their lives even if its effectiveness varies depending on individuals, their medical condition, race, sex, and age. That is why it is important to follow Fisher's Three Basic Principles when conducting such experiments. Without doing that, we may overlook the effectiveness of the drug, which may be masked by the noise component.

Q. You mean that it makes sense to follow Fisher's Three Basic Principles so that we do not miss a small effective factor hidden by the large noise component in agricultural and medical experiments, right? What about industrial experiments?

M. An important characteristic of industrial production is that it takes place in an artificially created space, i.e, the factory, which is not greatly influenced by the natural environment. A product that does not satisfy the standard is eliminated as a defect.

In the case of product design, the engineer is free to set the level of control factors. He can obtain an objective output only through the effect of the combination of levels he has chosen, and he must **eliminate or reduce (attenuate) the impact of noise components on the objective characteristic value of the output, until it is negligible.** These noise components are listed in *Table 2-13*.

Table 2-13. The Noise Component

Cause of noise component	
Noise from Production Environment	Noise from Market Environment
1. Changes in natural environment 2. Variability of the raw material 3. Changes in set condition	1. User handling 2. Time stability of the product

The product itself also requires stability against the noises it will encounter in the hands of the customer.

Fisher's Three Basic Principles are used in order to standardize deviation and variability, which may affect the experimental results, so that we can evaluate the error and avoid overlooking the small factor effect. However, for the engineer, **the combination of control factors, which he can freely choose, is the only method available for reducing or eliminating the influence of the changes in various noise factors.** The engineer must purposely consider all bias and variability. This is regarded as error (noise) from the viewpoint of Fisher's Three Basic Principles. However, **in industrial experiments, it is not necessary to introduce a test design based on repetition, randomization, or local control.**

Q. According to that concept, the test design will be very simple.

Not being **restricted by Fisher's Three Basic Principles means that the engineer is free of the constraints imposed on him by the statistician.** This means that statistics and experimental design that uses orthogonal arrays are independent, doesn't it?

M. Yes, indeed. The test itself does not have significance except as it can be used to determine the estimated factors for the optimum condition. It has been proven that selecting the optimum condition of levels that give a large factor effect will have sufficient reliability when the value is estimated as shown in *Table 3-11* and *Table 3-13*.

Q. Fisher's Three Basic Principles and test have been the engineer's Bible for a long time. **But the time has come to study the background and basis for these rules and to re-examine their applicability to industrial experimentation.**

M. Fisher's Three Basic Principles are still effective in medical and agricultural experiments, where noise is significant and a substantial effect can be expected even from a small factorial effect applied on a wide scale. In industrial experimentation, there are many reported examples of experiments that have yielded stable products, based on the new philosophy, in which error is positively introduced to make it interact with the main effects, or error factors are prepared and assigned outside of the orthogonal array. This results in highly reliable designs and contributes to corporate objectives and profit. It is also true that there are examples of applications that have had the wrong idea: that the number of experimental factors must be reduced, that the number of repetitions must be increased, or that the error of the experiment itself must be reduced. Such experiments actually fall short in providing solutions to industrial problems.

I have discussed error here because I wanted to emphasize that **the treatment of error is quite different in agricultural and industrial experimentation.**

Q. Why were Fisher's Three Basic Principles (which were based on agricultural experiments) even applied to industrial experimentation?

M. I think it was because no one sufficiently examined the difference between the contexts of industrial and agricultural experiments.

The application of experimental design to industrial experimentation has a 40 year history, but agricultural experiments have an even longer history. Therefore, it was natural to introduce a basic idea that had been successful in the field of agriculture to industry. But the factory itself is an artificial environment and the majority of phenomena there can be controlled or should be controlled. Otherwise the product will suffer from variability and this causes loss. In short, a different approach is required for treatment of noise (error).

Q. I think that only a few people have started thinking along those lines. Since Fisher's Three Basic Principles and test have a hundred year history, we should continue discussing your idea in order to find a better direction.

2-5. A Test Machine Experiment that Gives the Same Results as a Production Machine Experiment

Q. How can we obtain experimental results from a test machine (a small-scale experiment) that will correspond to the experimental results of a production machine?

When the test results of a small-scale experiment are transferred to large-scale equipment, the following often occurs:

1. The target values of the two situations do not coincide.
2. There are many defects.
3. The factorial effects obtained from the small-scale experiment differ for those of the large-scale equipment.
4. Changes in the environment of the large-scale equipment are directly reflected in product performance.

M. Regardless of whether he is involved in new product development or equipment and process design, **the engineer will always want to obtain experimental results from small-scale experiments that are flexible and inexpensive and that correspond to the large-scale equipment (actual production machines).** If he is able to do this, he can gain a great deal of useful information, in a short period of time, and with a small cost involvement. The optimum condition obtained from such small-scale experiments must not only satisfy the objective performance of the production machine, but also needs to be **a robust design that is stable against the noise effect** including changes in the environment.

Because **the data from the experimental machine are the only data available for new equipment design**, the objective of new equipment design research is to discover which kind of test should be done on the test machine.

We have an example in which the results of a one one-hundredth test machine experiment were used to complete the production machine specifications and condition settings even before the production machine was completed. But in order to achieve this, we need to study the new philosophy.

Q. The test machine should be used only for confirmation of the mechanism of the object of development and for examination of basic performance. This is true because the test machine is smaller than the production machine and often has a crude structure for temporary use. I do not think that a test machine can be used to estimate what will happen with a production machine. Nor can a test machine be used to examine the process control of the actual production machine. Moreover, we cannot obtain important information from a small-scale test machine. I do not think that results obtained from small-scale test equipment will hold for large-scale equipment (i.e. real machines).

M. **The belief that we cannot learn anything with small-scale equipment is completely wrong. Instead, we should believe we can do almost anything because it is small-scale equipment.** Because it is small, experimental costs are low and it is easy to reform and manipulate the equipment.

The reasons that results from small-scale equipment do not hold are:

1. The number of factors taken for small-scale equipment is too small.
2. The changes in environmental factors for large-scale equipment are large.
3. The appropriate experimental method is not used.

Q. Are you recommending that we do many factor experiments with small-scale equipment? Small-scale equipment does not have stability for condition setting because it is small. It usually gives different experimental results and this confuses us. We tend to use the one-factor-at-a-time approach so as to avoid such confusion. In order to raise so-called reproducibility, we use small factor experiments and one-factor-at-a-time (precision) experiments.

M. You have the wrong idea. **Small-scale equipment is test equipment** and is a tool for obtaining the information necessary for design. Because it does not use production equipment, it does not necessarily have stability.

What is important here is knowing how to design the production equipment, as well as how to set the optimum process condition. If the experimental results from the small-scale equipment, which we can control easily, lack reproducibility, then real equipment will give a much different output with a great deal of variability. Many defects will be produced.

Q. How can we obtain design information efficiently?

M. Two key points are important:

1. The factors that can be controlled in the actual machine, and
2. The levels of those factors that will make the objective value and the output coincide while remaining robust against changes in the environment, raw materials, and other noise components.

We need to find which factors have a strong relationship with the objective characteristic value. The multi-factor layout can be used if it does not require too much time and money, But because we have to make decisions in a short period of time, **it is better to assign main effects to all of the columns of the orthogonal array. It is not necessary to assign interactions and error terms.**

The objective performance should not vary much in the face of large changes in the environment or in raw materials in the actual machine. In order to select such levels more efficiently, we can transform our data into what is called the **Signal-to-Noise (S/N) ratio**. A larger S/N ratio is preferred.

$$S/N \text{ ratio} = \frac{\text{Signal component}}{\text{Noise component}}$$

Q. Can we believe that levels with a large S/N ratio will have a noise component, such as changes in the environment and raw materials, and can therefore be used to reduce the impact of such noise factors on the objective characteristic value?

M. Yes, we can. We can first determine the optimum condition by placing priority on the levels for that the S/N ratio is large, and then adjust other factors which have a weaker relationship with the S/N ratio, so as to achieve the objective level of the characteristic value.

It would also be helpful to check the **parameter-tolerance design**. Because each factor is more or less related to the objective-characteristic value, we need to consider trade-offs.

Q. I understand the importance of level selection for achieving stable products. But can't we obtain the relationship between each factor and characteristic value from a one-factor-at-a-time experiment?

M. Even though the results obtained from the one-factor-at-a-time experiment appear to be organized, they do not provide information that makes it possible to reproduce the results and reduce variability. This is because the background condition is fixed. Because the raw materials and environmental conditions change a great deal in the actual machine, a determination of the optimum condition based on a one-factor-at-a-time experiment is unsuitable.

What we should do is judge the effect of each factor on the objective characteristic value and the strength and weakness of factors against noise. This should be based on experimental results using the orthogonal arrays. We should control only those factors that have a strong relationship. We do not need to control factors with weak effects, and we can set those factors at any level we choose.

Because we need to achieve the objective value and to reduce variability through selection of the best combination of these factors, **we need many controllable factors.** This works as a trump card. And the use of orthogonal arrays is the only way to find these trump-card factors.

The normal condition and the worst condition for noise (about two-levels) need to be assigned to an outer orthogonal array in order to obtain the S/N ratio. These outside levels can be set by mixing noise factors. When a computer is used, we can use large arrays such as the L_{18} and the L_{36}.

When we perform evaluation experiments, we do not need to cover all combinations. Representative errant conditions (about two) and the standard one will do. The limits of the levels of controllable factors of the inner orthogonal array must be set widely, even to the point of producing samples that are not functional or, in the case of an experimental machine, to the extent of exceeding the actual machine capacity. Ideally we should use L_{18}, L_{36}, and L_{12} orthogonal arrays.

3

Classification of Quality Characteristics and Determination of Optimum Condition

3-1. Intro' ıction

When we determine optimum levels, it is usually better to use transformed data rather than raw data. Although measured (observed or unprocessed) raw data contain the original information from the experiments, there are many cases where we fail to derive appropriate optimum levels from the raw data if we do not process it. Using the example shown in *Table 2-9a*, this chapter explains the methods of raw data transformation and optimum condition determination suitable for meeting our experimental objectives.

This example concerns an electronic part that is partially plated with precious metal after the entire part is plated with chrome. *Figure 3-1* shows a sketch of this electronic part.

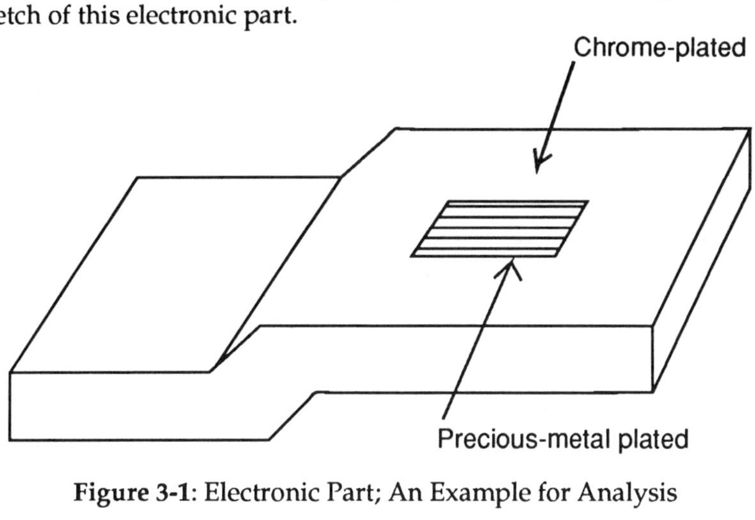

Figure 3-1: Electronic Part; An Example for Analysis

The experiment is expected to provide data regarding six characteristic target values (values are given in parentheses):

1. Wear rate of the chrome-plating film (less than 10 mg, more than - 20 db)
2. Film thickness of chrome plating where the precious metal is plated (8 ~ 12μ, more than - 6 db)
3. Adhesiveness between the base plate and the chrome plating film (more than 6 g, more than 15.6 db)
4. Glossiness of the chrome plating film (less than 60)
5. Thickness uniformity in the chrome plating film over the entire part (better-than-medium class)
6. Appearance of the part as a whole (better-than-good class).

The optimum condition of the experiment given in *Table 2-9* is determined on the basis of these six characteristic values.

3-2. Characteristic Value Classification Based on Data Form and Objective

The data that engineers handle can be classified as discrete or continuous data.

A **continuous value** is a quantity measured on a continuous scale, including fractions, whereas a **discrete value** is expressed as an integer, e.g., 1, 2, ...; it does not have fractional values. Qualitative data and ordered data are examples of discrete values. The representative measurement method giving qualitative data is the sensory test. This test compares a sample against the boundary sample or stepwise-quality-grade samples, so as to determine to which class the sample belongs. The result is expressed as a fixed marginal enumerative value. If the continuous value is grouped into concrete target levels, it may be processed as a multi-fractional value in the final analysis. Characteristic values can first be sorted into discrete and continuous values, and then classified into six subgroups. These subgroups are shown in *Table 3-1*.

On the other hand, characteristic data can also be classified on the basis of objectives. Continuous data can be grouped into **Smaller-the-Better (STB) data, Nominal-the-Best (NTB) data**, and **Larger-the-Better (LTB) data**.

STB data are measurements taken when our objective is to make the value as small as possible. Examples include impurity, carbon monoxide density, and the quantity of residue. NTB data are measurements taken when there is a predetermined target value. Examples include home electrical outlet voltage (100V), the thickness of a material, the degree of polymerization, and wire diameter. LTB data are measurements taken when our objective is to make the value as large as possible. Examples include material strength, adhesiveness, anti-weatherability, and electrical voltage resistance.

Data for all of these characteristics (STB, NTB, and LTB) can be transformed into the Signal-to-Noise (S/N) ratio, which has the property that the larger its value, the better the characteristic will be (i.e., the more stable) regardless of the original data type.

Table 3-1. Characteristic Value Classification

No.	Classification	Examples	
1	Simple discrete value	Countable data	Number of pinholes
2	Simple continuous value	Measurable data	Length, weight, strength, speed
3	Classified attribute value	• Classified data • Scaled-out data • Gauged data • Ordered data	Best, good, poor, etc. classification by comparison of test samples to standard
4	Classified continuous value	• Distribution with unknown total no. • Continuous curve data • Pure multi-fractional value	Powder, polymer; limit divided into classes; no order of division
5	Multi-enumerative value	The no. of samples in each class is counted, and the total no. is analyzed	No. of pinholes in different classes
6	Multi-variable value	Each sample is measured and classified, and the total is analyzed	Yield of different classes of agricultural products

3-3. Data Transformation

Although there are some exceptional cases where raw data obtained from an experiment can be used directly to derive the optimum condition without being processed, data transformation usually provides characteristic values that better suit the experimental objective.

In this section, the raw data shown in *Table 2-9* will be transformed for optimum condition determination, except for the data for glossiness, which is STB data. The glossiness data set is treated as an ordinary continuous value. This is the type of data most frequently used in traditional experimental design.

3-3-1. Smaller-the-Better Data

The wear rate in *Table 2-9* is expressed as STB data, which we prefer to be as small as possible. Non-negative STB data, $y_1, y_2, \dots, y_n$, are transformed into the S/N ratio according to the formula below. The S/N ratio is designated as η and is expressed in decibels (db).

$$\text{S/N ratio} \quad \eta = -10\log\frac{1}{n}(y_1^2 + y_2^2 + \cdots + y_n^2) \quad [\text{db}]$$

The smaller value of y_i, the larger the value of η.

The preceding formula is applied to transform data from experiments No. 1 – No. 8. It gives the following:

$$\text{No. 1} \quad \eta_1 = -10\log\frac{1}{3}(8^2 + 7^2 + 9^2)$$

$$= -18.107 \quad [\text{db}]$$

$$\text{No. 2} \quad \eta_2 = -10\log\frac{1}{3}\left(9^2 + 7^2 + 10^2\right)$$

$$= -18.846 \quad [\text{db}]$$

$$\text{No. 3} \quad \eta_3 = -13.680 \quad [\text{db}]$$

$$\text{No. 4} \quad \eta_4 = -8.653 \quad [\text{db}]$$

$$\text{No. 5} \quad \eta_5 = -6.690 \quad [\text{db}]$$

$$\text{No. 6} \quad \eta_6 = -29.965 \quad [\text{db}]$$

$$\text{No. 7} \quad \eta_7 = -26.659 \quad [\text{db}]$$

$$\text{No. 8} \quad \eta_8 = -23.741 \quad [\text{db}]$$

3-3-2. Nominal-the-Best Data

Chrome plating film thickness is expressed as NTB data. The target value is 10μ. The best method for obtaining this target thickness is to perform parameter design: First determine the condition that gives the smallest variability, and then adjust the volume of electrical current so as to obtain 10μ thickness. The optimum condition in this case is determined by calculating the difference between each measured value and the target characteristic value, 10μ. The S/N ratio is given by the following formula:

$$\eta = -10\log\frac{1}{n}\{(y_1 - y)^2 + (y_2 - y)^2 + \cdots + (y_n - y)^2\} \quad [\text{db}]$$

where $y_1, y_2, \dots$ yn are the data and y is the target characteristic value.

The smaller the difference between the data and the target value, the larger the value of η will be. The S/N ratio for each experiment is calculated as follows:

$$\text{No.1} \quad \eta_1 = -10\log\frac{1}{5}\{(9.3-10.0)^2+(8.3-10.0)^2+(8.9-10.0)^2$$
$$+(11.7-10.0)^2+(12.5-10.0)^2\}$$
$$= -4.387 \quad [\text{db}]$$

$$\text{No.2} \quad \eta_2 = -10\log\frac{1}{5}\{(10.6-10.0)^2+(9.5-10.0)^2+(10.3-10.0)^2$$
$$+(10.0-10.0)^2+(11.8-10.0)^2\}$$
$$= 1.305 \quad [\text{db}]$$

$$\text{No.3} \quad \eta_3 = -5.761 \quad [\text{db}]$$
$$\text{No.4} \quad \eta_4 = -6.557 \quad [\text{db}]$$
$$\text{No.5} \quad \eta_5 = -4.259 \quad [\text{db}]$$
$$\text{No.6} \quad \eta_6 = -9.182 \quad [\text{db}]$$
$$\text{No.7} \quad \eta_7 = -9.338 \quad [\text{db}]$$
$$\text{No.8} \quad \eta_8 = -4.657 \quad [\text{db}]$$

3-3-3. Larger-the-Better Data

Adhesiveness is an example of LTB data; therefore, values should be as large as possible. Data $y_1, y_2, \ldots, y_n$ are transformed into the S/N ratio according to the following formula:

$$\eta = -10\log\frac{1}{n}\left(\frac{1}{y_1^2}+\frac{1}{y_2^2}+\cdots+\frac{1}{y_n^2}\right) \quad [\text{db}]$$

The larger the value y_i, the greater the η will be. The S/N ratios for the adhesiveness data are calculated as follows:

$$\text{No.1} \quad \eta_1 = -10\log\frac{1}{2}\left(\frac{1}{10^2}+\frac{1}{15^2}\right)$$
$$= 21.413 \quad [\text{db}]$$

$$\text{No.2} \quad \eta_2 = -10\log\frac{1}{2}\left(\frac{1}{12^2}+\frac{1}{9^2}\right)$$
$$= 20.157 \quad [\text{db}]$$

$$\text{No.3} \quad \eta_3 = 11.584 \quad [\text{db}]$$
$$\text{No.4} \quad \eta_4 = 20.157 \quad [\text{db}]$$
$$\text{No.5} \quad \eta_5 = 15.199 \quad [\text{db}]$$
$$\text{No.6} \quad \eta_6 = 13.455 \quad [\text{db}]$$
$$\text{No.7} \quad \eta_7 = 16.635 \quad [\text{db}]$$
$$\text{No.8} \quad \eta_8 = 21.220 \quad [\text{db}]$$

3-3-4. Classified Data

The thickness uniformity of plated film is judged. We first create good, medium, and inferior classes. Then we count how many samples out of five fall into each class. The number of samples in each class is called the **frequency**. Whether a certain level is good or inferior is judged on the basis of the number of samples in Class I (good) and Class II (good plus medium). In order to obtain the number of samples in Class II, we need to add the number of samples in the medium class to the number of samples in the good class. This procedure is called **accumulation**. The following table summarizes **cumulative frequency** data:

No.	**Frequency**			**Cumulative Frequency**		
	G	*M*	*B*	I	II	III
1	4	1	0	4	5	5
2	1	3	1	1	4	5
.	.	.	.	.	.	.
.	.	.	.	.	.	.
.	.	.	.	.	.	.
8	0	0	5	0	0	5

No. 1 Class I = number of good samples = 4
Class II = number of good samples + number of medium samples
= 4 + 1 = 5
Class III = number of good samples + number of medium samples
+ number of inferior samples
= 4 + 1 + 0 = 5

No. 2 Class I = number of good samples = 1
Class II = number of good samples + number of medium samples
= 1 + 3 = 4
Class III = number of good samples + number of medium samples
+ number of inferior samples
= 1 + 3 + 1 = 5

The cumulative frequency for the rest of the experiments is derived similarly. Because the cumulative frequency of the last class is always the same, the last class is not used in determining the optimum levels.

3-3-5. Simple Discrete Values

The surface appearance of chrome plating can be judged good or inferior. The data generated will be simple discrete values. Data that are classified into only two classes are also called **[0, 1] data**. An accumulative frequency is also derived for these data by adding the frequency of the individual classes. The following table is an example of such a cumulative frequency:

No.	Frequency		Cumulative Frequency	
	G	B	I	II
1	4	1	4	5
2	5	0	5	5
.	.	.	.	.
.	.	.	.	.
.	.	.	.	.
8	0	5	0	5

No. 1 Class I = number of good samples = 4
Class II = number of good samples + number of inferior samples
= 4 + 1 = 5

No. 2 Class I = number of good samples = 5
Class II = number of good samples + number of inferior samples
= 5 + 0 = 5

Cumulative frequencies for the rest of the experiments can be derived similarly.

This is all of the data transformation necessary for the example in *Table 2-9*. Results are summarized in *Table* 3-2. The transformed data shown in *Table 3-2* are then used to calculate the process average and to draw a factorial-effects chart, from which we can determine the optimum level of each factor.

3-4. Experimental Objectives and Data Transformation

Engineering experimental objectives can be summarized as follows:

1. Adjust the output average to the target value (the NTB data).
2. Reduce variability.
3. Achieve (1) and (2) at the same time.
4. Reduce the output level as much as possible (STB data).
5. Increase the output level as much as possible (LTB data).
6. Reduce the number of defects.
7. Expand the stable region.

In general, most characteristic values will correspond to one of these objectives. The following explains the appropriate methods of data transformation for each of these experimental objectives:

3-4-1. Data Transformation when the Output Needs To Be Adjusted To The Target Value (NTB Data)

If there is only one factor that **linearly relates to the mean,** and if there is **no difference in the variability of this factor** in the entire region, then this factor can be used to reduce the difference between the output average and the target value. If the output is linearly related to the factor level, as it is in this case, the impact of the factor level itself on the output (inner noise) cannot be

Table 3-2: Chrome Plating Experiment with Transformation Data

$L_8\ (2^7)$ Orthogonal Array								Wear Rate [db]	Film Thickness [db]	Transformed Data							
Factor / Col.	*A* 1	*B* 2	*C* 3	*D* 4	*E* 5	*F* 6	*G* 7			Adhesiveness [db]	Glossiness	Thickness Uniformity I	Thickness Uniformity II	Thickness Uniformity III	Appearance I	Appearance II
1	1	1	1	1	1	1	1	−18.107	−4.387	21.413	52	4	5	5	4	5
2	1	1	1	2	2	2	2	−18.846	1.035	20.157	51	1	4	5	5	5
3	1	2	2	1	1	2	2	−13.680	−5.761	11.584	25	1	1	5	5	5
4	1	2	2	2	2	1	1	−8.653	−6.557	20.157	34	0	3	5	2	5
5	2	1	2	1	2	1	2	−6.690	−4.259	15.199	69	0	0	5	2	5
6	2	1	2	2	1	2	1	−29.956	−9.182	13.455	77	1	3	5	0	5
7	2	2	1	1	2	2	1	−26.659	−9.338	16.635	86	2	3	5	2	5
8	2	2	1	2	1	1	2	−23.741	−4.657	21.220	95	0	0	5	0	5

eliminated, even though this factor is effective for adjusting the output level. If there is another factor, whose level is **non-linearly related to the output**, then the impact on the output from the changes in the level of this factor can be attenuated. This is known as the nonlinear theorem for stable design.

Nonlinear Theorem for Stable Design (from *Reliability Design Example Handbook* by the Japanese Standard Association)

Suppose the relationship between the final characteristic value, y, and factors that compose the system (e.g., the parts and material size and their characteristic value) is as shown in *Figure* 3-2. Here, the variability of the final characteristic value, y, is smaller when parameter A changes from A_1 to A_2 compared to the change from A_3 to A_4.

Therefore, it is obvious that it is better to chose the level of parameter A between A_3 and A_4 as long as there exists some method that allows for achievement of the target value.

The difference between the target value and y, i.e., distance "a" in the figure, which will be generated when the value of A is set between A_3 to A_4, can be eliminated by arranging the level of parameter B that has a linear relationship to the final characteristic value, y.

In order to determine the level of each parameter, we need to study many factors, and this makes it important to use the experimental design method that effectively uses orthogonal arrays and analyzes the data using S/N ratios.

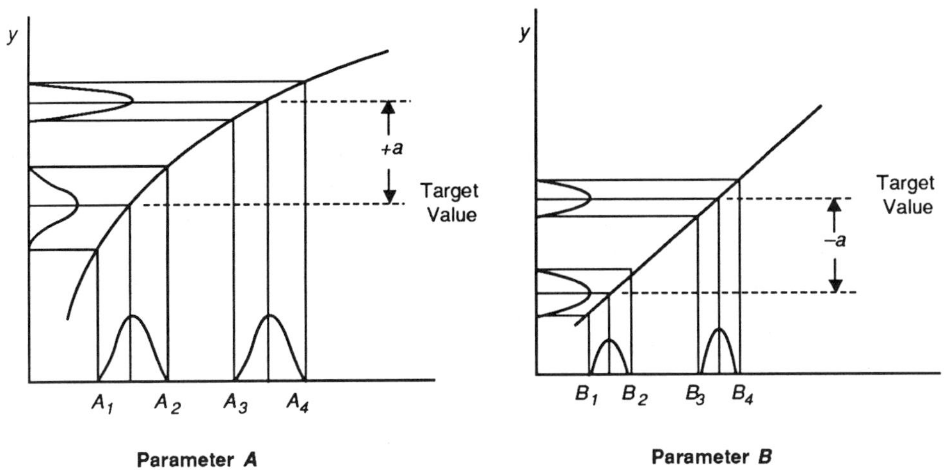

Figure 3-2: An Approach to Parameter Level Determination

When many factors that are linearly related to the output exist, it is often difficult to decide which factor should be used to adjust the output. If we arbitrarily select a factor to adjust the output without considering inner and outer noise, we may end up having many defects and excessive output, which requires other experiments.

If there is no factor that can be used to adjust the output continuously, the following formula may be used:

$$\eta = -10\log\frac{1}{n}\{(y_1 - y)^2 + (y_2 - y)^2 + \cdots + (y_n - y)^2\} \quad [\text{db}]$$

where $y_1, y_2, \ldots, y_n$ are the data and y is the target characteristic value.

In this case, deviation from the target value and variability are evaluated simultaneously. This situation is shown in *Figure* 3-3. In this case, it is necessary to include an adjustment factor as one of the experimental factors, which can be used to adjust the output average to the target value.

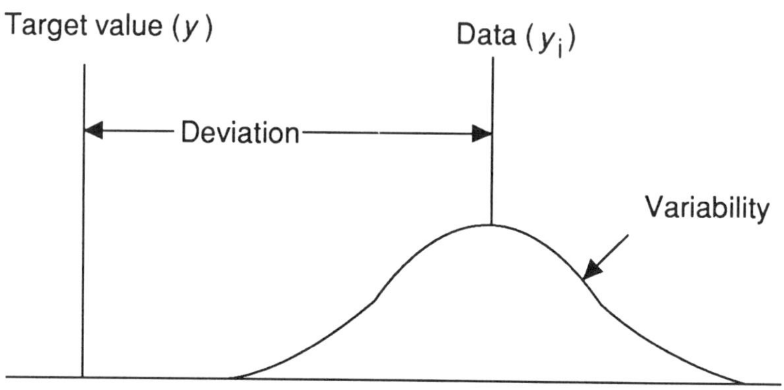

Figure 3-3: Deviation from the Target Value and Variability

The value of the S/N ratio (η) becomes large as the difference between the target value and the data becomes small. The combination of levels that gives the smallest variability can be obtained by using the S/N ratio (η) to determine the optimum condition.

We have explained optimization in the case of continuous values. The following discussion explains classified data. **If data are given as continuous values rather than classified data, it is possible to perform a more appropriate analysis (i.e., one that suits the experimental purpose).** However, the continuous data sometimes needs to be treated as classified data. For example, electrical current leakages cannot be treated as a continuous value if there is an infinitely large amount of leakage. In such cases, it is necessary to treat the data as classified data by dividing the level of leakage into different levels. **In experiments involving data from unpolymerized synthetic reactions, a motor that does not rotate, etc., we have to treat the data as classified data instead of being able to use continuous values.**

For example, if a semi-glossy surface is needed for a chloroethylene sheet, the classes are determined as in *Table 3-3*. The class is derived on the basis of the level of satisfaction, i.e., quite satisfactory (O), acceptable (Δ), and unacceptable (×). If the classification is based on the appearance mode, the letters I, II, III, IV, and V can be used. Whichever classification is used, level A_1 is the better level because, according to *Table 3-3*, it gives more semi-glossy samples.

Table 3-3: Appearance Evaluation of Chloroethylene Sheet Based on Classified Data

Appearance Judgment	Glossy	Interme-diate	Semi-Glossy	Interme-diate	Satin Finish
Satisfaction Ranking	X	Δ	O	Δ	X
Ranking Mode	I	II	III	IV	V
A_1	0 (0)	1 (1)	6 (7)	2 (9)	1 (10)
A_2	3 (3)	4 (7)	2 (9)	1 (10)	0 (10)

( ) Cumulative frequency

3-4-2. Data Transformation when Variability Needs To Be Reduced

When it is necessary to reduce variability, different approaches can be taken depending on whether we can change the central value. There are three possibilities:

1. The central value of variability cannot be changed.
2. The central value of variability can be moved arbitrarily regardless of the factor that creates the variability.
3. A factor in the experiment is used to reduce the variability, as well as to adjust the central value to the target value.

Because there is no factor that can be used for adjustment in Case (1), the method described in Section 3-4-1 should be used to simultaneously improve deviation from the target and variability.

Because the adjustment factor is independent in Case (2), variability alone needs to be optimized first, and the actual value can be adjusted to the target value later using an adjustment factor. The data are transformed according to the following formula:

$$\eta = -10\log\frac{1}{n}\{(y_1-\bar{y})^2+(y_2-\bar{y})^2+\cdots+(y_n-\bar{y})^2\}$$

where $y_1, y_2, \dots, y_n$ are the data.

$\bar{y}$ in the equation is the average. If the data are closer to the average, the S/N ratio becomes a large number.

A similar procedure is frequently used when making a hole with a drill. In such a case, the best procedure is to first determine the optimum condition and to then move the hole to the best position using the adjustment factor.

3-4-3. Data Transformation When Variability Needs To Be Reduced and the Output Needs To Be Adjusted to the Target Value

In this case, it is better to design the tolerance after designing the parameters. The factor used in the experiment must relate both to the output (target characteristic) and to variability.

The analysis procedure involves obtaining the S/N ratio, η, and the general mean, S_m, from the outer data, which are assigned with mixed error factors, accelerated tests, etc.

The following items need to be calculated if the data are given as $y_1, y_2, \dots, y_n$:

$$ST = y_1^2 + y_2^2 + \cdots + y_n^2$$

$$SM = \frac{1}{n}(y_1 + y_2 + \cdots + y_n)^2$$

$$S_e = ST - SM$$

$$V_e = \frac{S_e}{(n-1)}$$

$$\text{S/N ratio:} \quad \eta = 10 \log \frac{\frac{1}{n}(SM - V_e)}{V_e}$$

General mean: $S_m = 10 \log SM$

The larger η is, the smaller the variability will be. The larger S_m is, the larger the output becomes. The procedure can be summarized as follows:

1. Choose the combination of levels that gives the smallest variability based on η. The output level derived from this step is usually different from the target value.
2. Adjust the difference between the output level derived from (1) and the target value using the adjustment factor (i.e., the factor that gives large changes in S_m but does not influence η).

It is difficult to find the adjustment factor using a small orthogonal array. The use of a larger array, such as an L_{36} is usually recommended. An L_{36} or an L_{18} must be used because these arrays distribute interactions, more or less evenly, to each column.

Parameter design does not cause a cost increase because it reduces variability and adjusts the output level using a combination of factor levels of the cheapest parts and processes. If sufficient reduction of variability cannot be achieved through parameter design, each part and process must be adjusted so as to reduce the amplitude of the output level from the central value of the system and unit control condition. In this case, because control capability is increased, such adjustments involve the use of high-class parts and higher grade control equipment. This will increase the cost. This procedure is called tolerance design.

Tolerance design determines which parts and processes need to be modified and to what degree it is necessary to increase their control capacity.

Tolerance design is done through experiments in which the factors used in parameter design or factors involved in the current control system are assigned to an L_{18} or an L_{36}. The ideal levels in this case would be:

First level: $m-\sqrt{3/2}\sigma$

Second level: m

Third level: $m-\sqrt{3/2}\sigma$

σ: standard deviation of the variability of the level

m: central value of the parameter

Factor variation in the experimental results is separated into two components: primary variation effects and secondary variation effects, before carrying out analysis of variance.

Primary effect: the linear component of factor level effects on the output.

Secondary effect: the nonlinear component of factorial effects, which becomes the noise component, regardless of the changes in factor levels.

Figure 3-4 shows the concept of primary and secondary effects.

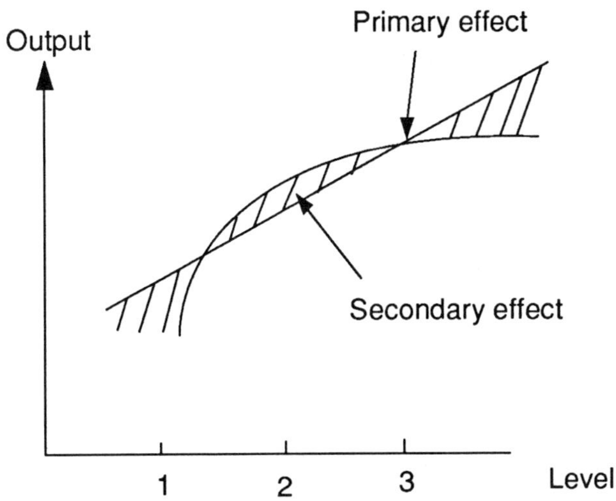

Figure 3-4: Primary and Secondary Effect of a Factor

A key point here is that we should **raise the control capability of the factor level with a large primary effect**. Suppose variance, V_T, and contribution, ρ, are obtained from the results of a tolerance design experiment.

For example, if the primary effect (*l*) of *B*, *C*, and *D* is large, a part that gives one-half error (variability) is used for part *B*, and parts which give one-tenth error are used for parts *C* and *D*, then the new error variance, V_e, becomes:

$$V_e = V_T\left[\left(\frac{1}{2}\right)^2 \rho_{Bl} + \left(\frac{1}{10}\right)^2 (\rho_{Cl} + \rho_{Dl}) + \rho_e\right]$$

Here, it is necesary to minimize productivity, P, which is the sum of the loss, *L*, due to the increased quality and upgrade cost, *C*.

$P = L + C$

For details, see the explanation of tolerance design in Chapter 9.

In the previous example, a continuous value was used. However, even with classified data, we may have a certain target class. In such a case, data need to be arranged as in *Table 3-3* in order to select the levels that allow the largest number of samples to fall into the desired classification mode. The same idea as that used in parameter design is applied to determine those levels. *Figure 3-5* is a representation of the data in *Table 3-3*.

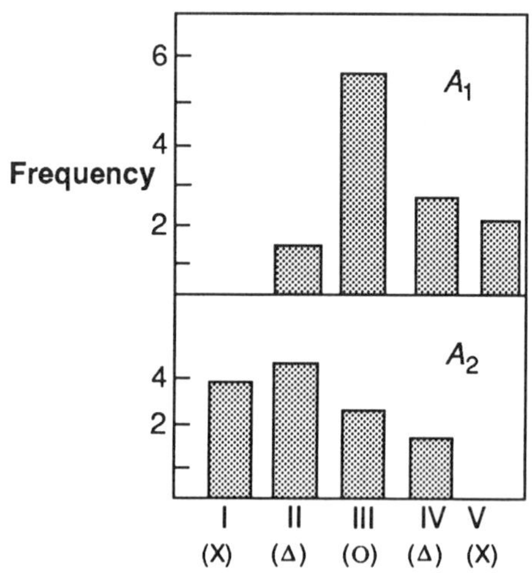

Figure 3-5: Frequency Distribution of A_1 and A_2

If the class marked O is desirable, level A_1 must be selected. When the combination of levels is determined using this method, the distribution of the output will be sharply concentrated around Class III, which is marked with O. It is thus possible to achieve an output level at the target value with small variability, even if we use classified data.

3-4-4. Data Transformation when the Output Level Needs To Be as Small as Possible (STB Data)

Examples include carbon monoxide contained in exhaust gas, residue, impurity, etc. **For continuous values,** data should be closer to zero, even if the output value is non-negative and suffers from variability. When temperature is the characteristic value, we can use an absolute temperature. Observations often give a negative value, which by definition is impossible. In such a case, we can approximate the negative data as zero. Data transformation uses the following formula:

$$\eta = -10\log\frac{1}{n}(y_1^2 + y_2^2 + \cdots + y_n^2) \quad \text{[db]}$$

where $y_1, y_2, \ldots, y_n$ are the data.

As the value approaches zero, η becomes large.

If the data are classified, count the samples in each class and choose the level that gives the largest number. From the data shown in *Table 3-4*, level A_2 must be chosen as the best level.

Table 3-4: Classified Data

Judgment	**Small**	**Medium**	**Large**
Level of Satisfaction	O	Δ	X
A_1	2 (2)	7 (9)	1 (10)
A_2	6 (6)	4 (10)	0 (10)

( ): Cumulative frequency

3-4-5. Data Transformation when the Output Needs to be as Large as Possible (LTB Data)

Material strength, adhesiveness, and durability are examples of this type of data. Data transformation is as follows:

$$\eta = -10\log\frac{1}{n}\left(\frac{1}{y_1^2} + \frac{1}{y_2^2} + \cdots + \frac{1}{y_n^2}\right)$$

where $y_1, y_2, \ldots, y_n$, are continuous data.

As the value increases, η becomes large. When we test for rupture strength, some samples may not be destroyed. The data can then be classified as strong, medium, and weak, and the level that gives the largest number of samples in the strong class should be chosen. From the classified data in *Table 3-5*, level A_2 must be chosen as the best level.

Table 3-5: Classified Data

Judgment	Small	Medium	Large
Level of Satisfaction	O	D	X
A_1	2 (2)	6 (8)	2 (10)
A_2	6 (6)	3 (9)	1 (10)

( ) Cumulative frequency

3-4-6. Data Handling when the Number of Defects Needs To Be Reduced

A defective product is one that does not satisfy the standard. If the current production line produces many defects, the best way to reduce them is to eliminate the causes of variability existing in the process. These factors often come from raw materials and working conditions. In some companies, the only improvement methods accepted are the use of higher grade parts and improvements in process control capacity. Replacement of parts of a similar grade and changes in condition settings are not accepted. Other companies accept only one method, which corresponds to tolerance design, because the impact of a change in the level of one factor on other characteristic values cannot be confirmed on the production line. But it is not difficult to reduce the number of non-functioning products, i.e, defects produced in the production line, at the design stage. This is true because **a design that is robust against inner noise (level changes) and outer noise (effect of error factors caused by the environment and raw materials) can be achieved by performing parameter design and tolerance design at the design stage.**

3-4-7. Data Handling when the Stable Region Needs to be Expanded

Stability has two aspects: functional region expansion and stability. In the case of photocopying (xerography), the former corresponds to the electrostatic charge, which needs to be very large. From the data given in Table 3-6, level A_1 is the better level. In this case, the level that gives a large electrostatic charge band can be obtained by analyzing the lower value of the picture forming electrostatic charge band as STB data and the upper value as LTB data. The direct method can be used. But, calculation is complicated and the method takes more time. Optimum levels can be selected by drawing a factorial effects chart without doing analysis of variance.

Table 3-6: Electrostatic Charge Band of Xerographic Copying and Picture Formation

Electro-static Charge	400	450	500	550	600	650	700	750	800
A_1	X	O	O	O	O	O	O	O	X
A_2	X	X	O	O	O	X	X	X	X

O = Good
X = Bad

With light bulbs, consumers prefer a longer life product. The results of a durability test on light bulbs with two different production conditions is summarized in *Table 3-7*. In this example, A_1 is the better level. **Minute accumulation analysis** is another durability test analysis method. This method is effective **when we want to find the declining pattern (distribution) of durability**. Because it is desirable to have longer durability, it can be treated as an LTB characteristic.

Table 3-7: Light Bulb Durability Test

Test Hour	0	500	1000	1500	2000	2500	3000	3500	4000
A_1	O	O	O	O	O	O	O	O	X
A_2	O	O	O	O	X	X	X	X	X

O = Good
X = Bad

We have explained several data transformation methods for different experimental objectives. Different methods may be used to achieve the same objective. In such a case, **analysis using continuous data is always preferred to analysis with classified data.** The data transformation that gives the value closer to the estimate obtained from the orthogonal array experiment should be applied.

3-5. The Level Sum, Process Average, and Factorial Effects Chart

The best levels of each factor must be chosen in order to determine the optimum condition. A simple and reliable method for achieving this is to derive the process average from the level sum and then draw a factorial effects chart.

3-5-1. Level Sum Calculation

Before drawing a factorial effects chart, we must first calculate the level sum. We have to be careful here that the sign indicating the level sum is the same as the sign of the level itself. **The level sum is the sum of the experimental data for a particular level of a factor.**

For example, in the case of the wear rate shown in *Table 3-2*, the level sums of level A_1 of factor A and level G_2 of factor G are as follows:

The level sum of the first level, A_1, of factor A is:

A_1 = the sum of experimental data for level A_1
$= (-18.107) + (-18.846) + (-13.680) + (-8.653)$
$= -59.286$

The level sum of the second level, G_2, of factor G is:

G_2 = the sum of experimental data for level G_2
$= (-18.846) + (-13.680) + (-6.690) + (-23.741)$
$= -62.957$

For classified data, the level sum is obtained by adding the frequency of each set. For example, the level sums of level A_1 of factor A and level G_2 of factor G, for the thickness uniformity of plated metal, are calculated as follows:

The level sum of the first level, A_1, of factor A is:

Class I $A_1 = 4 + 1 + 1 + 0 = 6$
Class II $A_1 = 5 + 4 + 1 + 3 = 13$
Class III $A_1 = 5 + 5 + 5 + 5 = 20$

The level sum of the second level, G_2, of factor G is:

Class I $G_2 = 1 + 1 + 0 + 0 = 2$
Class II $G_2 = 4 + 1 + 0 + 0 = 5$
Class III $G_2 = 5 + 5 + 5 + 5 = 20$

Other level sums are similarly obtained. The results are summarized in *Table 3-8*.

3-5-2. Process-Average Calculation

The level sum is the sum of all experimental data for a particular level. We need to determine the average effect of the level before drawing our factorial effects chart. **The average effect of a level is calculated by dividing the level sum by the number of data involved in calculating the level sum.** The average effect of the level is called the process average.

Let us calculate the process average of the first level of factor A and the second level of factor G, in the case of wear rate. The process average of the level is indicated by a bar sign over the level name (e.g. $\overline{A}$, etc.).

The level sums of A_1 and G_2 for wear rate involved four pieces of data. The average effect of a factor is derived by dividing the level sum by the number of data:

Table 3-8: Level Sum (Supplementary Table)

Factor	Level	Wear Rate [db]	Film Thickness (db]	Adhesive-ness [db]	Glossi-ness	Thickness Uniformity of Plated Film I	II	III	Appearance I	II
A	1	–59.286	–15.671	73.311	162	6	13	20	16	20
	2	–87.046	–27.436	66.509	327	3	6	20	4	20
B	1	–73.599	–16.794	70.224	249	6	12	20	11	20
	2	–72.733	–26.313	69.596	240	3	7	20	9	20
C	1	–87.353	–17.348	79.425	284	7	12	20	11	20
	2	–58.979	–25.759	60.395	205	2	7	20	9	20
D	1	–65.136	–23.745	64.831	232	7	9	20	13	20
	2	-81. 196	-19.362	74.989	257	2	10	20	7	20
E	1	–85.484	–23.987	67.672	249	6	9	20	9	20
	2	–60.848	–19.120	72.148	240	3	10	20	11	20
F	1	–57.191	–19.860	77.989	250	4	8	20	8	20
	2	–89.141	–23.247	61.831	239	5	11	20	12	20
G	1	–83.375	–29.464	71.660	249	7	14	20	8	20
	2	–62.957	–13.643	68.160	240	2	5	20	12	20
Total Sum	(T)	–146.332	–43.107	139.820	489	9	19	40	20	40

$$\text{Process mean of } \overline{A}_1 = \frac{-59.286}{4} = -14.822$$

$$\text{Process mean of } \overline{G}_2 = \frac{-62.957}{4} = -15.739$$

The process averages for the other factors are calculated similarly. **For classified data, the process average is expressed in terms of a percentage by setting the last class at 100%**. For example, the process average of A_1 and G_2, for thickness uniformity of the plated film, are calculated as follows:

The process average of A_1 is:

$$\text{Class I } \overline{A}_1 = \frac{6}{20} \times 100 = 30(\%)$$

$$\text{Class II } \overline{A}_1 = \frac{13}{20} \times 100 = 65(\%)$$

$$\text{Class III } \overline{A}_1 = \frac{20}{20} \times 100 = 100(\%)$$

The process average of G_2 is:

$$\text{Class I } \overline{G}_2 = \frac{2}{20} \times 100 = 10(\%)$$

$$\text{Class II } \overline{G}_2 = \frac{5}{20} \times 100 = 25(\%)$$

$$\text{Class III } \overline{G}_2 = \frac{20}{20} \times 100 = 100(\%)$$

The results are summarized in *Table 3-9*.

Table 3-9: Process Average for Classified Data

Class / Level	Cumulative Frequency			Process Mean(%)		
	I	II	III	I	II	III
A_1	6	13	20	30	65	100
G_2	2	5	20	10	25	100

All of the process averages are summarized in *Table 3-10*. Continuous values in the table have two digits below zero due to a computer output restriction.

Table 3-10: Process Average (Supplementary Table)

Factor	Level	Wear Rate [db]	Film Thickness [db]	Adhesive-ness [db]	Glossi-ness	Thickness Uniformity of Plated Film			Appearance %	
						I	II	III	I	II
A	1	–14.82	–3.92	18.33	40.50	30	65	100	80	100
	2	–21.76	–6.86	16.63	81.75	15	30	100	20	100
B	1	–18.40	–4.20	17.56	62.25	30	60	100	55	100
	2	–18.18	–6.58	17.40	60.00	15	35	100	45	100
C	1	–21.84	–4.34	19.86	71.00	35	60	100	55	100
	2	–14.74	–6.44	15.10	51.25	10	35	100	45	100
D	1	–16.28	–5.94	16.21	58.00	35	45	100	65	100
	2	–20.30	–4.84	18.75	64.25	10	50	100	35	100
E	1	–21.37	–6.00	16.92	62.25	30	45	100	45	100
	2	–15.21	–4.78	18.04	60.00	15	50	100	55	100
F	1	–14.30	–4.97	19.50	62.50	20	40	100	40	100
	2	–22.29	–5.81	15.46	59.75	25	55	100	60	100
G	1	–20.84	–7.37	17.92	62.25	35	70	100	40	100
	2	–15.74	–3.41	17.04	60.00	10	25	100	60	100
Average	($\overline{T}$)	–18.29	–5.39	17.48	61.125	22.5	47.5	100	50	100

3-5-3. Drawing the Factorial Effects Chart

The **factorial effects chart** shows the relationship between the factor and the output on a diagram. It helps us to select the best levels. The factor levels are indicated on the horizontal axis (X axis). Points are marked at the intersection of each factor level and the process average. The process average is indicated on the vertical axis (Y axis). Points for two levels of each factor are connected with a straight line. The higher the process average, the larger the S/N ratio will be.

The **process average for classified data** is generally shown as a **vertical bar chart** in which the white portion indicates good results, the shaded portion indicates medium results, and the black portion indicates inferior results. A level giving a larger portion of the white bar is usually taken as better.

Figure 3-6 is a factorial effects chart in which all characteristic values are indicated on the X axis. This total factorial effects chart makes the relationship among characteristic values clearly visible and enables us to make an efficient adjustment of levels (trade-off, balancing) when we determine the optimum condition.

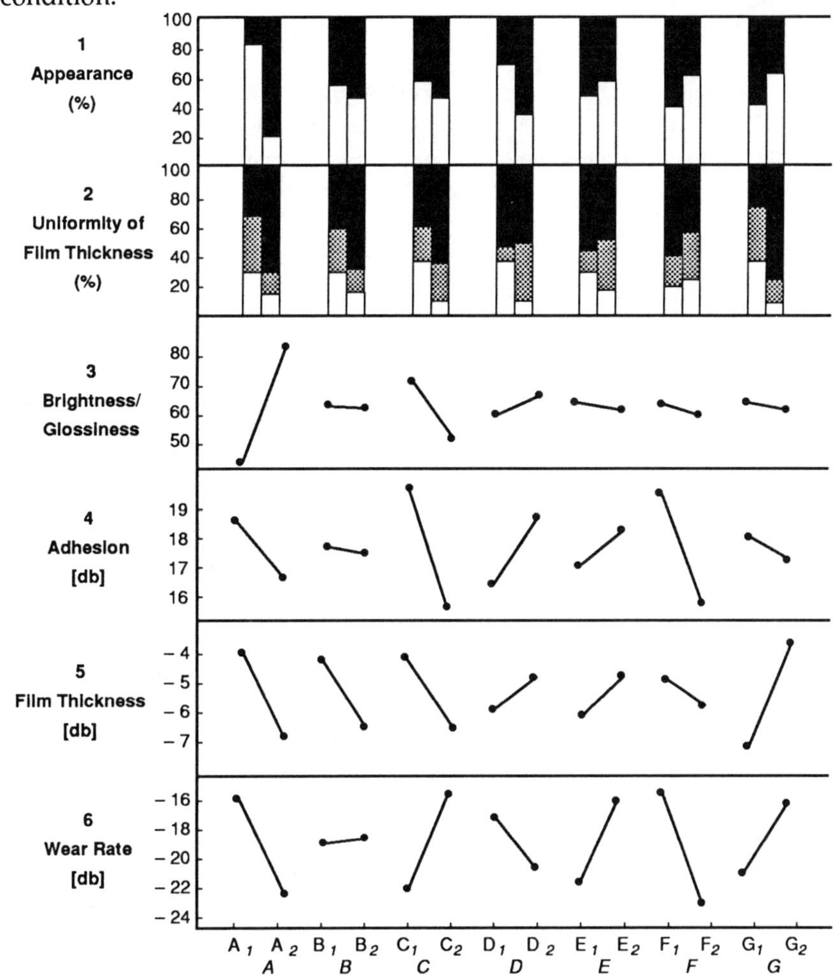

Figure 3-6: Total Factor Effect Chart

3-6. Determination of the Optimum Condition and Its Estimate

Ideally, we want to have the same combination of optimum levels to achieve each characteristic value and for the product as a whole. In many cases, however, the optimum condition for each characteristic value is different. This section explains the derivation of the optimum condition for each characteristic value and the derivation of an estimate of characteristic value improvement when the current condition is replaced by the optimum condition.

The **estimate of optimum condition ($\hat{\mu}$)** is derived by adding the gains derived from the selection of the levels of particularly effective factors to the mean of the experimental data ($\overline{T}$). In general,

Estimate of the optimum condition:

$\hat{\mu} = \overline{T}$ + sum of the differences between the total mean and each process average, for each level of particularly effective factors

This equation is used to estimate the optimum condition ($\hat{\mu}$). **Particularly effective factors** are limited to approximately between one-third and one-half of the total factors.

There are four methods for obtaining estimates of the optimum condition:

1. **The all-difference-accepting method**

   With this method, every factor used in the experiment is accepted as effective, and the difference between every process average and the total mean is added to obtain the estimate.

2. **The shrinkage factor method**

   With this method, if the ratio (factor variance, V)/(error variance) is greater than one, the difference between the process average and the total mean ($\overline{T}$) is added after multiplying it by the shrinkage factor, β.

$$\beta = \begin{cases} 0 & \Rightarrow F_0 = \dfrac{V}{V_e} \le 1 \\ 1 - \dfrac{1}{F_0} & \Rightarrow F_0 = \dfrac{V}{V_e} > 1 \end{cases}$$

   This method is the most reliable, but it is necessary to calculate the variation and the variance.

3. **The F test method**

   With this method, an Analysis of Variance (ANOVA) table is created and estimation is based only on the levels of statistically significant factors. This method requires us to conduct an analysis of variance.

4. **Estimation using only strong effects**

   With this method, factors with a large effect are taken from the factorial effects chart, and estimation is based on the levels of these factors. This method does not require analysis of variance. It is the simplest method to use. Even though reliability of estimation is not high, the actual estimation is quite similar to that derived from the β coefficient and F test methods.

Dr. Genichi Taguchi has compared these methods based on actual experiments. Estimates based on these four methods and the differences between the estimates and the actual values are listed in *Table 3-11*. In our example, the fourth method provided an estimate that was exactly the same as the actual value. *Table 3-13* gives estimation results based on the fourth method, where strong effect factors were selected from the factorial effects chart. It compares the size of factorial effects and the current condition, indicated as experiment No. 1. The result seems to have sufficient reliability. We can thus conclude that **estimation using the fourth method is quite acceptable for practical use.** The greatest advantage of this method is that it does not require analysis of variance.

Table 3-11: Various Estimation Methods and Their Error

No.	Method for Estimating the Optimum Condition	Estimated Value	Actual Value	Error
1	The all-difference-accepting method	5.1	4.1	+1.0
2	The β coefficient method	4.0	4.1	–0.1
3	The F test method	4.7	4.1	+0.6
4	Use only large effect factors method	4.1	4.1	±0

Taguchi, System of Experimental Design 3d ed., ASI Press, p. 570.

In the next section, estimation of the optimum condition will be based on the fourth method.

3-6-1. Estimation of the Optimum Condition for Continuous Data

This section describes determination of the optimum condition and its estimate when the data are continuous, e.g., wear rate, film thickness, adhesiveness, and glossiness.

Wear rate

Effective factors for wear rate are *A*, *C*, *E*, and *F*, and their optimum condition levels are A_1, C_2, E_2, and F_1. Data with a large S/N ratio should be used.

Estimate of the optimum condition:

$$\hat{\mu}(A_1C_2E_2F_1) = \overline{T} + (\overline{A}_1 - \overline{T}) + (\overline{C}_2 - \overline{T}) + (\overline{E}_2 - \overline{T}) + (\overline{F}_1 - \overline{T})$$

$$= \overline{A}_1 + \overline{C}_2 + \overline{E}_2 + \overline{F}_1 - 3 \times \overline{T}$$

$$= -14.82 - 14.74 - 15.21 - 14.30 - 3 \times (-18.29)$$

$$= -4.20\,[\text{db}] = 1.622\ (\text{mg})$$

Film thickness

Effective factors for film thickness are *A*, *B*, *C*, and *G* and their optimum levels are A_1, B_1, C_1, and G_2.

Estimate of the optimum condition:

$$\begin{aligned}\hat{\mu}(A_1B_1C_1G_2) &= \overline{A}_1 + \overline{B}_1 + \overline{C}_1 + \overline{G}_2 - 3\times\overline{T} \\ &= -3.92 - 4.20 - 4.34 - 3.41 - 3\times(-5.39) \\ &= +0.30\,[\mathrm{db}] = 10 \pm 0.97 = 9.03 \sim 10.97\,(\mu)\end{aligned}$$

Adhesiveness

Effective factors for adhesiveness are *C*, *D*, and *F* and their optimum levels are C_1, D_2, and F_1.

Estimate of the optimum condition:

$$\begin{aligned}\hat{\mu}(C_1D_2F_1) &= \overline{C}_1 + \overline{D}_2 + \overline{F}_1 - 2\times\overline{T} \\ &= 19.86 + 18.75 + 19.50 - 2\times 17.48 \\ &= 23.15\,[\mathrm{db}] = 14.371\,(\mathrm{g})\end{aligned}$$

Glossiness

A lower value of glossiness is preferred. The effective factors are *A* and *C* and their optimum levels are A_1 and C_2.

Estimate of the optimum condition

$$\begin{aligned}\hat{\mu}(A_1C_2) &= \overline{A}_1 + \overline{C}_2 - 1\times\overline{T} \\ &= 40.50 + 51.25 - 61.125 \\ &= 30.625\end{aligned}$$

3-6-2. Estimation of the Optimum Condition for Classified Data

This section discusses optimum condition estimation for classified data, such as thickness uniformity of plated film and appearance. The effective levels of classified data are determined by choosing those levels that give a longer white ("good") bar chart in the factorial effects chart. If there is no difference between the levels, in terms of the size of the white portion, the shaded ("medium") portion is added to the white portion to compare the levels. In order to estimate the optimum condition, the omega (Logit) transformed value of percentage (%) is used, because percentage data does not have additivity. The **omega transformation** is given as follows:

$$-10\log\left(\frac{1}{P}-1\right)[\mathrm{db}]$$

where *P* is the percentage.

Appendix *Table 2* can be used to find the **omega value** instead of performing the calculation.

Thickness Uniformity of Plated Film

The effective factors for thickness uniformity of plated film are *A, B, C,* and *G*, and their optimum levels are A_1, B_1, C_1, and G_1.

Estimate of the optimum condition:

Class I $$\begin{aligned}\hat{\mu}(A_1B_1C_1G_1) &= \overline{A}_1 + \overline{B}_1 + \overline{C}_1 + \overline{G}_1 - 3\times\overline{T} \\ &= \overline{A}_1(30\%) + \overline{B}_1(30\%) + \overline{C}_1(35\%) + \overline{G}_1(35\%) - 3\times\overline{T}(22.5\%) \\ &= -3.679 - 3.679 - 2.687 - 2.687 - 3\times(-5.370) \\ &= +3.378 \text{ [db]} \\ &= 68.5\,(\%)\end{aligned}$$

Class II $$\begin{aligned}\hat{\mu}(A_1B_1C_1G_1) &= \overline{A}_1 + \overline{B}_1 + \overline{C}_1 + \overline{G}_1 - 3\times\overline{T} \\ &= \overline{A}_1(60\%) + \overline{B}_1(60\%) + \overline{C}_1(60\%) + \overline{G}_1(70\%) - 3\times\overline{T}(47.5\%) \\ &= 1.761 + 1.761 + 1.761 + 3.680 - 3\times(-0.434) \\ &= 10.265 \text{ [db]} \\ &= 91.4\,(\%)\end{aligned}$$

It is not necessary to estimate the last class, because it is 100%.

Appearance Judgment

The effective factors for appearance improvement are *A, D, F,* and *G* and their optimum levels are A_1, D_1, F_2, and G_2.

Estimate of the optimum condition

Class I $$\begin{aligned}\hat{\mu}(A_1D_1F_2G_2) &= \overline{A}_1 + \overline{D}_1 + \overline{F}_2 + \overline{G}_2 - 3\times\overline{T} \\ &= \overline{A}_1(80\%) + \overline{D}_1(65\%) + \overline{F}_2(60\%) + \overline{G}_2(60\%) \\ &\quad -3\times\overline{T}(50\%) \\ &= 6.021 + 2.688 + 1.761 + 1.761 - 3\times 0 \\ &= 12.231 \text{ [db]} \\ &= 94.35\,(\%)\end{aligned}$$

It is not necessary to estimate the last class because it is 100%. If the classified data give 0% or 100%, the omega transformation gives $-\infty$ and $+\infty$, which are not good estimates. Therefore, 0% and 100% need to be specially treated in the following way:

0% case — Consider it as 1/(number of data) and omega transform it. In this example, the number of data is 40, thus, it will be: $(1/40)\times 100 = 2.5\ \% = -15.911$ [db].

100% case — Consider it as (number of data - 1)/(number of data) and omega transform it. In this example, the number of data is 40, thus, it will be: $[(40 - 1)/40]\times 100 = 97.5\ \% = +15.911$ [db].

The levels chosen for each characteristic value are listed in *Table 3-12*.

Table 3-12: Effective Factors, Best Levels, and Optimum Condition

Factor	Level	Wear Rate (db)	Film Thickness (db)	Adhesiveness (db)	Glossiness	Thickness Uniformity of Plated Film	Appearance	Optimum Condition		
								Priority on Quality	Priority on Productivity	Current Condition
A	1	○	○		◎	○	◎	●	●	●
	2									
B	1		○			○		●	●	●
	2									
C	1		○	◎		○		●	●	●
	2	○			○					
D	1						○	●	●	●
	2			○						
E	1									●
	2	○						●	●	
F	1	○		◎				●		●
	2						△		●	
G	1					○				●
	2		◎				△	●	●	
Estimate		–4.20 [db] (1.622 mg)	+0.3 [db] (9.03 ~ 10.97 μ)	23.15 [db] (14.371g)	30.625	Class I 68.5% Class II 91.4%	Class I 94.35%			
Target		>–20 [db] (> 10 mg.)	>–6 [db] (8~12 μ)	>15.5 [db] (> 6 g)	< 60	Better-than-med. class	Better-than-good class			

3-7. Total Evaluation to Determine the Final Optimum Condition

In order to determine the final optimum condition, adjustment (trade-off, balancing) among characteristic values is necessary, because the choice of levels may conflict between characteristic values. By totally evaluating *Table 3-12*, we obtain the following final optimum condition:

A_1, B_1, C_1, D_1, E_2, F_1, and G_2.

In addition to quality, the cost and productivity of the product must be evaluated before we finally decide on the optimum condition. Examining various cases to compare the relationship between each optimum condition and the target characteristic values is recommended.

Condition (A_1, B_1, C_1, D_1, E_2, F_2, and G_2), which replaces the level of F with F_2, may be selected as another candidate for the final optimum condition because it increases the output by 1.5 times. Estimation is also done for the current condition (which corresponds to experiment No. 1). *Table 3-13* shows the estimates of each characteristic value for these two optimum condition candidates, as well as the estimate for the current condition.

Wear rate

$$\hat{\eta}(A_1C_2E_2F_1) = -14.82 - 21.84 - 15.21 - 14.30 - 3\times(-18.29)$$
$$= -11.30\,[\text{db}] = 3.673\,(\text{mg})$$
$$\hat{\eta}(\text{A}_1\text{C}_1\text{E}_2\text{F}_2) = -14.82 - 21.84 - 15.21 - 22.29 - 3\times(-18.29)$$
$$= -19.29\,[\text{db}] = 9.215\,(\text{mg})$$
$$\hat{\eta}(\text{A}_1\text{C}_1\text{E}_1\text{F}_1) = -14.82 - 21.84 - 21.37 - 14.30 - 3\times(-18.29)$$
$$= -17.46\,[\text{db}] = 7.464\,(\text{mg})$$

Film thickness

$$\hat{\eta}(A_1B_1C_1G_2) = -3.92 - 4.20 - 4.34 - 3.41 - 3\times(-5.39)$$
$$= +0.3\,[\text{db}] = 10\pm 0.97 = 9.03 \sim 10.97\,(\mu)$$
$$\hat{\eta}(A_1B_1C_1G_1) = -3.92 - 4.20 - 4.34 - 7.37 - 3\times(-5.39)$$
$$= -3.66\,[\text{db}] = 10\pm 1.52 = 8.48 \sim 11.52\,(\mu)$$

Adhesivenes

$$\hat{\eta}(C_1D_1F_1) = 19.86 + 16.21 + 19.50 - 2\times 17.48 = 20.61\,[\text{db}]$$
$$= 10.728\,(\text{g})$$
$$\hat{\eta}(C_1D_1F_2) = 19.86 + 16.21 + 15.46 - 2\times 17.48 = 16.57\,[\text{db}]$$
$$= 6.738\,(\text{g})$$

Table 3-13: Estimates of Each Characteristic Value for Two Optimum Condition Candidates

No.	Optimum Condition	Reason	Wear [db]	Film Rate [db]	Adhesive-ness [db]	Glossi-ness [db]	Thickness Uniformity of Plated Film (%)	Appear-ance (%)
1	$A_1B_1C_1D_1$ $E_2F_1G_2$	Priority on Quality	–11.30 (3.673)	–0.3 (8.96 ~ 11.04)	20.61 (10.728)	50.375	I 31.0 II 65.3	I 88.1
2	$A_1B_1C_1D_1$ $E_2F_2G_2$	Priority on Productivity	–19.29 (9.215)	–0.3 (8.96 ~ 11.04)	16.57 (6.738)	50.375	I 31.0 II 65.3	I 94.4
Current Condition	$A_1B_1C_1D_1$ $E_1F_1G_1$	Current Condition	–17.46 (7.464)	–3.66 (8.48 ~ 11.52)	20.61 (10.728)	50.375	I 68.5 II 92.9	I 76.8
Experiment No. 1 (Current Condition)		Expected Value	18.387 (7 ~ 9)	–4.387 (8.3 ~ 12.5)	21.413 (10 ~ 15)	52	I 80 II 100	I 80
		Target Value	> –20 [db] (< 10 mg)	> –6 [db] (8 ~ 12 μ)	> 15.5 [db] > 6 g	< 60	Better than medium class	Better than good class

Glossiness

$$\hat{\mu}(A_1C_1) = 40.50 + 71.0 - 61.125 = 50.375$$

Thickness Uniformity of Plated Film

Class I
$$\begin{aligned}\hat{\mu}(A_1B_1C_1G_2) &= \overline{A}_1(30\%) + \overline{B}_1(30\%) + \overline{C}_1(35\%) + \overline{G}_2(10\%) \\ &\quad -3 \times \overline{T}(22.5\%) \\ &= -3.679 - 3.609 - 2.687 - 9.541 - 3 \times (-5.370) \\ &= -3.476\,[\text{db}] \\ &= 31.0(\%)\end{aligned}$$

Class II
$$\begin{aligned}\hat{\mu}(A_1B_1C_1G_2) &= \overline{A}_1(65\%) + \overline{B}_1(60\%) + \overline{C}_1(60\%) + \overline{G}_2(25\%) \\ &\quad -3 \times \overline{T}(47.5\%) \\ &= 2.688 + 1.761 + 1.761 - 4.770 - 3 \times (-0.434) \\ &= 2.742\,[\text{db}] \\ &= 65.3\,(\%)\end{aligned}$$

Class I
$$\begin{aligned}\hat{\mu}(A_1B_1C_1G_1) &= \overline{A}_1(30\%) + \overline{B}_1(30\%) + \overline{C}_1(35\%) + \overline{G}_1(35\%) \\ &\quad -3 \times \overline{T}(22.5\%) \\ &= -3.679 - 3.679 - 2.687 - 2.687 - 3 \times (-5.370) \\ &= 3.378\,[\text{db}] = 68.5\,(\%)\end{aligned}$$

Class II
$$\begin{aligned}\hat{\mu}(A_1B_1C_1G_1) &= \overline{A}_1(65\%) + \overline{B}_1(60\%) + \overline{C}_1(60\%) + \overline{G}_1(70\%) \\ &\quad -3 \times \overline{T}(47.5\%) \\ &= 2.688 + 1.761 + 1.761 + 3.680 - 3 \times (-0.434) \\ &= 11.192\,[\text{db}] \\ &= 92.9\,(\%)\end{aligned}$$

Appearance

Class I
$$\begin{aligned}\hat{\mu}(A_1D_1F_1G_2) &= \overline{A}_1(80\%) + \overline{D}_1(65\%) + \overline{F}_1(40\%) + \overline{G}_2(60\%) \\ &\quad -3 \times \overline{T}(50\%) \\ &= 6.021 + 2.688 - 1.760 + 1.760 - 3 \times 0 \\ &= 8.709\,[\text{db}] = 88.1\,(\%)\end{aligned}$$

$$\begin{aligned} \text{Class I} \quad \hat{\mu}(A_1D_1F_2G_2) &= \overline{A}_1(80\%) + \overline{D}_1(65\%) + \overline{F}_2(60\%) + \overline{G}_2(60\%) \\ &\quad -3 \times \overline{T}(50\%) \\ &= 6.021 + 2.688 + 1.760 + 1.760 - 3 \times 0 \\ &= 12.229\,[\text{db}] = 94.4\ (\%) \end{aligned}$$

Note: The rounding is different from the calculation given in the previous calculation.

$$\begin{aligned} \text{Class I} \quad \hat{\mu}(A_1D_1F_1G_1) &= \overline{A}_1(80\%) + \overline{D}_1(65\%) + \overline{F}_1(40\%) + \overline{G}_1(40\%) \\ &\quad -3 \times \overline{T}(50\%) \\ &= 6.021 + 2.688 - 1.760 - 1.760 - 3 \times 0 \\ &= 5.189\,[\text{db}] = 76.8\ (\%) \end{aligned}$$

Optimum condition (1) gives improved wear rate and film thickness while it lowers the current level of plated-film-thickness uniformity.

Optimum condition (2) improves the film thickness and appearance, but it lowers the adhesiveness and the thickness uniformity of the plated film. However, both optimum conditions clear the target characteristic values. **The choice between these two will depend on the situation of new product development at competing companies and the company's productive capability.**

A comparison of the estimate of the current condition and its actual performance, as indicated by the data for experiment No. 1, shows a good match. This shows that estimation of the optimum combination of factor levels based only on strong-effect factors has proven to be sufficiently accurate for our purposes.

4

Experimental Design Application Manual

Chapter 3 covered the effectiveness and application of experimental design, including assignment of factors to orthogonal arrays and optimum condition determination.

This chapter introduces an experimental design application manual that includes orthogonal array assignment, procedures for data analysis, and an application checklist (DM sheet). Even beginners can achieve a degree of success comparable to that obtained by trained engineers if they follow this checklist. The basic approach (philosophy) regarding items cited in the checklist will be explained in Chapter 5.

The process of experimental design can be divided into the five stages summarized in *Table 4-1*. Contents of the checklist are summarized in *Table 4-2* through *Table 4-6*. These are included not only to demonstrate how to use orthogonal arrays and to analyze experimental results, but also to aid in the smooth execution of development themes. A step-by-step description is related to the flow of the application of orthogonal arrays and data analysis. The DM sheet (*Table 4–7*) can be used to summarize the reader's own experimental design method.

Table 4-1: Summary of Experimental Design Application

No.	Stage	Activity Content
1	Preparation	• Determine the experimental objective and target-characteristic values. • Determine whether it is worthwhile to carry out the experiment under the current conditions.
2	Determination of factors and levels	• List all factors that relate to the objective. • Select and determine factors and levels to be tested.
3	Assignment	• Assign factors and levels to the orthogonal array. • Determine the outer condition (noise).
4	Experiment	• Eliminate obstacles hindering the experiment.
5	Data analysis	• Determine good levels through data transformation and calculation of level sums and process averages. • Determine the optimum condition and calculate its estimate.

Table 4-2: Experimental Design Application: First Stage—Preparation

Preparation	Activity Content	Checklist	OK	No
1	List items relating to the development theme (e.g., new product, design, equipment design, process design) that have attracted attention.	• Have there been changes in the development theme from new-product developments at other companies? • Have changes in market structure made the development theme obsolete? • Will the product be competitive in the face of advancements in developing countries? • Are there changes in export market conditions due to monetary strength? • Do political/economic conditions in foreign countries require revision of the development theme?		
2	List methods for achieving the development theme.	• List any method as long as there is a possibility it may work (even if it has never been tried). • Evaluate and rank the suggested methods in terms of importance and necessary function. • Have you used brainstorming and the KT method? (Did you use methods that stimulate thinking?)		
3	Select the method that shows the most potential for achieving the development theme.	• Are the target values of the development theme expressed in concrete numbers? • Are the other targets (e.g., characteristic values of products supplied by other companies) listed and understood by everyone? • Is the target authorized? Has it been presented to the engineers? • Are the targets divided into the absolute target and desired target? • Is the project flexible enough to all market/economic changes (on an everyday basis).		

Table 4-2 (Cont'd): Experimental Design Application: First Stage—Preparation

Preparation	Activity Content	Checklist	OK	No
		• Do the targets include items such as cost, productivity, and operational simplicity?		
4	Collect information on the method selected for the experiment.	• Are there restrictions on the technology? • Estimate the cost, time, and no. of engineers required for technological development. Compare that estimate with the existing conditions. • Has information on competitive products been collected? • Have engineers previously in charge been interviewed? Did you collect data from the past? • Have you reviewed rival company patent information?		

Table 4-3: Experimental Design Application:
Second Stage—Determination of Factors and Levels

Factor and Level Determination	Activity Content	Checklist	OK	No
1	List more than 50 factors related to the development theme. (Step 1)	• Have you gathered everyone's input? • Did you use information from previous patents, articles, engineers, and other experts? • Did you include even those factors eliminated by the previous person in charge? • Have you identified many new ideas? • Have you picked up factors from discussion with a wide range of people? • Is there a surprise factor based on new ideas (e.g., brainstorming)? • Did you include factors suggested by superiors, other engineers, and subordinates? • Did you visit the production site and laboratory to discuss the factors? • Were those who stubbornly maintain rigid beliefs and refuse to change excluded from the discussion? • Were discussions continued until about 50 factors were listed?		
2	Classify the factors listed into control factors, indicative factors, block factors, supplementary factors, and noise factors. (Step 2)	• Did you list more than 30 control factors? • Did you continue the search when the no. of control factors was less than half the factors listed? • Are you sure that an important factor was not eliminated by the person in charge based on personal opinion? • Is the no. of untested, unknown (new) control factors about half of the control factors? If the no. is much less, search for additional new factors.		

Table 4-3 Cont'd: Experimental Design Application: Second Stage—Determination of Factors and Levels

Factor and Level Determination	Activity Content	Checklist	OK	No
3	Estimate the effect of control factors on target values through discussions with many engineers. Classify control factors into most promising, promising, and other. (Step 3)	• Did many people including the person in charge participate in the factor ranking? • When it is difficult to rank a factor, treat it as promising. • Did past experimental results bias the factor ranking? • Are more than half of the most promising factors new? If not, search for additional new factors and rank again.		
4	Set the levels of each factor. (Step 4)	• Do most factors have three levels? If the factorial effect is expected to be linear, two-levels can be used. • For three-level settings, set the second level at the target value. Set the first and third levels at values that will give nonfunctional samples. It is acceptable for one-third to one-half of the sample to be nonfunctional. • The levels chosen must have a possibility of achieving the development theme. Do not use levels that are obviously no good.		
5	Select levels that will achieve low cost and high productivity. (Step 5)	• Can the levels selected be expected to achieve low cost (e.g., low-cost parts, low-processing energy, low-priced medicine, and small-sized parts). • Can the levels selected be expected to achieve high productivity (e.g. shortened processing time, improved uniform quality, and reduced size and weight)? • Do the levels selected give light, thin, short, small products rather than heavy, thick, long, large products? • Determine appropriate levels by examining the limitations of the experimental equipment and the range of levels you will use. • Will many factors achieve the target cost?		

Table 4-4: Experimental Design Application: Third Stage—Assignment

Factor and Level Determination	Activity Content	Checklist	OK	No
1	Select the inner orthogonal array and assign factors and levels. (Step 6)	• Did you select an L_{12}, L_{18}, or L_{36} array? (An L_8, L_{16}, L_{32}, or L_{27} can also be used.) • Assign all the most promising factors to the array. If there are unused columns, assign other factors to them. • Interactions and error should not be assigned. Assign only main effects, even if interactions/error exist. • When using idle column, multi-level or nested factor methods, check other examples and corresponding levels again.		
2	Select the outer orthogonal array and assign factors and levels. (Step 7)	• There is no particular restriction imposed on the outer array if a computer is used for parameter design. • If the optimum condition is to be obtained from the experiment, compound the noise factors and use a layout for the outer array. • The compounding result should have a standard condition and two inferior standard conditions on either side of the standard. • Are there too many samples that come from the outer array (experiment unrealistic)? • Are factors with appropriate inner and outer noises compounded? • Is the compounding condition based on past performance? (e.g., Does the applied enforcement, durability, or promotion condition follow industrial standards?) • Are the levels of outside factors appropriate? Do they take into account the impact of the noise expected to come from large-sized production machines and the marketplace?		

Table 4-4 Cont'd: Experimental Design Application: Third Stage—Assignment

Factor and Level Determination	Activity Content	Checklist	OK	No
3	Develop instructions for conducting the experiment. (Step 8)	• Do the instructions give easy procedures? Prepare the instructions carefully so as to avoid operator mistakes. • Determine the order of experiments. When it is difficult to change factor levels, use the easiest experiment order. The designer should determine the order and keep a record. • If not confident about the progress of the experiment, suspend it and re-examine the factors and levels. • The inner array should only have control factors. Check it again. • Explain the instructions, in detail, to operators at the experiment site. If instructions are confusing, rewrite them. • Do not repeat the same experiment. • Compare and confirm the number of evaluation samples and test samples. • Is there any combination of levels that cannot be tested?		
4	Determine the data to be collected. (Step 9)	• Collect all characteristic-value data relating to the development theme. Prepare a column for recording data in the experimental instructions. • Classify characteristic values into main, sub, real, and alternative characteristics. Rank order them. • Use continuous data as much as possible. Try to transform classified and simple discrete data (this gives higher reliability and greater analytical power). • Is data collected so as to improve both the average and variability? • Does the collection of outer-array data allow for transformation into S/N ratios?		

Table 4-5: Experimental Design Application: Fourth Stage—Experimentation

Experiment	Activity Content	Checklist	OK	No
1	Prepare the environment for the experiment.	• Will the experiment be done with an experimental machine or the actual machine? An experimental machine allows a relatively large range of levels. With a production machine, levels must be restricted to avoid breakdown. • Check the experimental order. It can be set for convenience rather than for the improvement of probability. • If an experimental machine is used, is the difference between the experimental machine and the production machine used as a factor? Are appropriate levels selected? Were compounded levels of outer factors chosen against noises from production and the marketplace? • Is there an adequate rental period for equipment? Is the skill level of operators adequate? • Was the lot, quantity, and grade of raw material confirmed with the material supplier? • Is the supply of raw material sufficient to cover unsuccessful experiments? (Is additional material available?) • Is the material used in the experiment supplied by the same company as that used in production? Does it have the same grade? Is the cost, quality, and delivery time of the material as good? • Compare the experiment instructions with experimental equipment capacity. (Is the experiment to be run possible?) • What restrictions are imposed by people, measurement equipment, and test equipment?		

Table 4-5 Cont'd: Experimental Design Application: Fourth Stage—Experimentation

Experiment	Activity Content	Checklist	OK	No
		• Were sufficient personnel, measuring apparatus, and enforcement test equipment secured for sample evaluation?		
		• Have current conditions (control, type, etc.) been considered?		
2	Perform a final check before the experiment.	• Ask everyone involved, "If this experiment fails, what do you think will be the cause?"		
		• Revise the plan so that anticipated problems can be eliminated. Prepare countermeasures.		
		• Is the person in charge confident of good results? If not, revise the plan before implementation.		
		• Consider all possibilities in which the experimental objective might not be achieved (e.g., problems in the previous process, subsystems, or parts).		
3	Anticipate experimental results prior to the experimentation. Consider actions and countermeasures for different degrees of achieving the experimental objectives.	• Obtain concensus regarding the content of follow-up experiments.		
		• Will an additional or alternative experiment be done if the experimental results are not satisfactory? (Obtain concensus.)		
		• If additional experiments are expected, be sure that the content of the additional experiments can be confirmed within the current plan.		
		• Critically examine the content of the experiment (to the point where you are almost trying to disprove the possibility of achieving a concrete method and results for the target).		
		• Check to see if everyone knows the schedule for each case.		
		• Is alternative technology prepared? Are the levels evaluated properly?		

Table 4-6: Experimental Design Application: Fifth Stage—Data Analysis

Optimum Experimentation	Activity Content	Checklist	OK	No
1	Check the data. (Step 10)	• If data are missing, estimate them using the successive approximation method.		
2	Transform the data. (Step 11)	• Transform data into STB, LTB, and NTB S/N ratios. • Transform classified data and simple discrete data into cumulative data. • If the data set has infinitives, treat it as classified variable data and transform it into cumulative data.		
3	Calculate the level sum. (Step 12)	• Calculate the level sum for each factor level. The no. of data will be different in the case of multi-level and idle columns.		
4	Calculate the process average. (Step 13).	• Divide the level sum by the no. of data. Assume that the process average is 100% for the last class.		
5	Perform analysis of variance. (This step may be skipped.) (Step 14)	• Check to see if the variation between experiments is equal to the sum of the variation of assigned levels. • Complete the analysis of variance (ANOVA) table.		
6	Construct a factorial-effects chart. (Step 15)	• Construct a factorial-effects chart for all factors.		
7	Estimate the optimum condition. (Step 16)	• (When analysis of variance is not done) Choose one-third to one-half of all factors considered to be effective. Base the estimate of optimum condition on these. (When analysis of variance is done) Estimate the optimum condition using only significant factors. • List several optimum conditions from different perspectives, e.g., quality, cost, production, and loss. Apply the very best condition to the new product/production process by comparing their estimates with target characteristic values.		

Table 4-7. DM Sheet

related factor chart	Date __________, 19___
experiment target	expected effect large (◎) medium (○)
schedule	target
assignment	→ ☐
characteristic value main characteristic secondary characteristic	

5

Description of the Experimental Design Application Manual

This chapter explains the approach (philosophy) underlying the use of the checklists that provide the procedural order for the experiment. In most experimental design training, only the calculation methods are stressed. The approach that forms the basis for the experimental design is ignored, even though it is crucial for its success. The approach and concepts underlying each step are explained in a question-and-answer format.

5-1. Factors and Levels

Q. How can we locate which factors to test in the experiment?

M. The "**factor list**" locates which factors will achieve the development theme. **The combination of levels of listed factors must contain the "solution" required to achieve the development theme**. This means that the probability of accomplishing the objective increases as the number of factors listed increases.

The following are some key points in factor listing:

1. List many factors (more than 50).
   - Include as many people as possible that are related to the project (superiors, workers, researchers, producers, etc.) in the discussion.
   - Invite specialists, consultants, sales people, and project planners from different departments to participate in the discussion.
   - It is best to include even those people who always reject others' opinions, as well as those who do not echo traditional ways of thinking.
   - In the process of factor listing, do not try to criticize others' opinions. Rather, try to encourage others to come up with different kinds of factors.
   - Prepare samples, study other companies products, data, catalogs, parts, etc., so that the discussion can cover as many details as possible.

2. List many factors that are peculiar, unexpected, and of a kind that have not been listed before.
   - Welcome outsiders' ideas.
   - It is sometimes better to exclude people who were formerly in charge of the project from the discussion. This allows us to get away from stereotypical views. These points are particularly important for bringing about new phenomena and epoch-making improvements. The listing usually ends when we have 30 to 50 factors.

5-1-1. Factor Listing

Q. But isn't it a waste of time to list so many factors when it is practically impossible to do so many experiments? If we meet with a small number of people including the supervisor, we can list factors quickly without much disorder. We can then start the experiment quickly, because what needs to be done can be determined in a short period of time.

M. It is true that we can quickly determine what should be done without a lot of confusion if the number of people involved is restricted. However, discussions limited to related personnel have the following problems:

1. The approach taken by these people is biased, and the pattern of behavior and determination tends to be the same. It is therefore difficult to find a breakthrough that leads to the optimum condition.
2. Because any one individual will be shortsighted to a certain degree, we cannot justify views and conclusions about a project based only on one person's ideas.

Q. Well, I can understand the shortcomings of activities with a small number of people, but . . .

M. We are looking for an unexpected and dramatic result that will make it possible to fulfill the development theme. This cannot be achieved by people who only have very similar ideas. We need to have new ideas for the development of new products. People who are not so closely related to the experimental objective can often suggest such new ideas. This is why we need to effectively utilize outsiders who do not have a direct involvement with the project.

Q. I think it is better to list factors theoretically.

M. It is not at all necessary that factor listing be done theoretically. **Rather, factor listing should be intuitive.**

The theory itself does not provide any breakthroughs, because it is simply an explanation of our common sense and just describes a phenomenon mathematically. Since brainstorming must be intuitive, outside people can participate in factor listing. New results do not come from research that follows a theory. The theory itself only allows us to anticipate the limits. **An unexpected and novel idea or phenomenon comes only from going beyond a theory.**

Q. Theories must be highly respected because they express phenomena with mathematics.

M. Well, I do not disagree with you, but I have never heard of a case where a theory has lead to the development of a new product in industrial corporations. A theory can only be applied to a simple system. A product is a complex combination of many systems, and the output cannot be explained with a theory alone. If the theory could tell us the result, we would not need to do the experiment. The experiment is necessary because the theory cannot anticipate the results. The theory can be used as a point of departure for thinking, but it should not be used to restrict our ways of thinking.

Q. What is the advantage of listing many factors?

M. That is a good question, and in fact, there are two.

1. A larger number of factors gives a higher possibility of finding factors which will allow us to achieve the experimental objective (target value and reduced variability).
2. The involvement of related people in the factor listing process helps to achieve agreement about the direction and content of the development.

The first point means that because the factors are listed by all people related to the project, factors that allow us to achieve the target condition will not be missed. Furthermore, a discussion involving many people makes it possible to pick up many different kinds of factors for improving the results.

The second point is very important. Factor listing involving all related personnel has the following three advantages:

1. It motivates the entire development team.
2. It promotes cooperation in the conduct of the experiment.
3. It prevents any one person with a strong personality from dominating the discussion. This helps in the development of an unbiased and objective plan.

Q. I think these two points are related to management and are not directly related to the experimental design method.

M. The experimental design method itself is a technique, a set of technical skills, which is different from a management method. Because the experimental design method provides the optimum condition, the experiment should include the factors and their levels that will achieve the target. But it is also necessary to have the skills to guide the experiment to success and to motivate people. These are skills of management.

Q. This means that we cannot achieve our development target using only the knowledge of experimental design, doesn't it?

M. That is quite right. **Achieving the target depends on the engineer's ability to plan and prepare conditions before the experiment is run.**

If the factors and levels used in the experiment do not contain the "solution," the result will be failure. I think that factor listings that involve everyone is readily accepted in Japan because quality-control activities are popular there.

Q. I think it is easier for the engineer in charge to just list the factors and do the experiment.

M. Sure, it is easy and uncomplicated for one person to do whatever he wants. But what is important here is that even if the research theme is assigned to only one engineer, its success or failure affects the entire corporation. I do not mean to ignore what the person in charge thinks, but that person must become the promoter of target achievement by **motivating the entire project team and collecting everyone's knowledge** rather than by trying to solve the problem with only one's own technological abilities.

We should not and cannot expect a completely new product to develop solely from the engineer in charge because his knowledge will always be limited.

Q. But, you are asking engineers to become promoters? Engineers prefer to work by themselves.

M. I think engineers tend to have that inclination. The key to a successful experiment is to encourage researchers' creativity and to accumulate the wisdom and knowledge of others.

We often see that an engineer tries hard to explain why his experiment has failed. It is not easy to hear the reasons that an experiment failed. The major reason for most failed experiments is that the engineer did not take the necessary steps prior to running the experiment.

Q. You cannot always blame the engineer for an unsuccessful experiment.

M. The success of an experiment is determined by what it accomplished in terms of its objective. If an engineer repeats an experiment because he initially omitted factors, the failure is his fault. On the other hand, even if the experiment gives failing results, in terms of achieving the objective, it can still be considered a success if it helps to change the direction of the development policy. Sufficient factor listing is the first requirement of a successful experiment.

5-1-2. Factor Classification

Q. Why do we classify the factors?

M. Suppose an experiment was done based on the factors and levels listed in *Table 5-1*. In this case, C and D are completely meaningless factors.

Table 5-1: Syntheses Experiment

	Factor	Level 1	Level 2
A	Catalyzer	A_1	A_2
B	Temperature	60	70
C	Test time	Morning	Afternoon
D	Operator	Mr. A	Mr. B

Q. Why is that? I think we can judge which levels are better.

M. Well, we can judge which levels are better. But an important point is whether the information is useful for the engineer.

As for factor *D*, it is meaningless to check it unless either Mr. A or Mr. B is the only operator. Also, factor *C*, the difference in the time of day of operation, is irrelevant because it does not contribute to performance. Since we should select only those factors that are effective for improving performance, it is very important to classify the factors.

Q. That means that, in order to improve the efficiency of the experiment, we need to classify the factors and assign them properly.

M. We currently classify the factors into the following five categories:

1. **Control factors**: Factors that the engineer can freely choose without constraint (parts on a mechanical drawing, elements of an electric circuit, a process flow of chemical reaction, etc.).
2. **Indicative factors**: Factors that have levels, but for which it does not make sense to select the best level (condition of product usage; condition of enforcement and deterioration test; difference in product, person, or equipment, etc.).
3. **Block factors**: These are factors that have levels, but for which there is no technical implication for selecting the best level. (i.e., changes in the levels of those factors do not have a controlling effect, even for adjustment). They are selected to prevent their effects from being included with the effects of control factors. Block factors may include machines, test day, lot, etc..
4. **Supplementary factors**: Factors used to record the experimental condition and environment.
5. **Noise factors**: Factors other than those mentioned above whose condition is not recorded but which will influence the experimental results.

For parameter design, we distinguish the signal (adjustment) factor in addition to the ones previously listed:

6. **Signal (Adjustment) Factors**: Factors that do not influence the noise component (variability) and which can be used to adjust the output to the target value.

Q. Would you explain the classification and assignment of factors?

M. *Table* 5-2 shows the relationship between factors and corresponding orthogonal arrays. The control and block factors are assigned to the **inner orthogonal array,** and the indicative and noise factors are assigned to the **outer orthogonal array**. The outer orthogonal array has factors and levels that produce noise. When a mixed noise factor or an enforcement test condition is used, the outer orthogonal array is replaced with a one-way layout. The key point here is that **control factors are assigned to all columns of the inner orthogonal array**.

Table 5-2: The Orthogonal Array and Factor Assignment

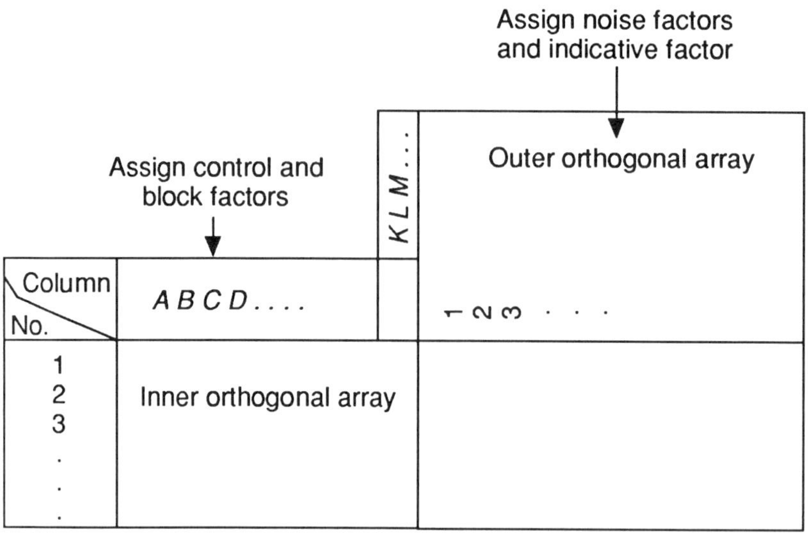

Q. I don't understand how to use the block factor.

M. The block factor is assigned to the inner orthogonal array. The following are examples of block factor assignment:

1. A tire example: In the development of a tire, four tires R_1, R_2, R_3, and R_4 are assigned as factor R to the inner orthogonal array (as a block factor) to prevent the difference in wear caused by tire position from being mixed with the effect of the control factor.
2. A measuring apparatus example: When the number of samples is large, two measuring apparatuses, C_1 and C_2, are used. In such a case, factor C is assigned to the inner orthogonal array to prevent the difference in measuring apparatus from being mixed with the effect of the control factor.

5-1-3. Factor Ranking

Q. Because the number of columns in an orthogonal array is limited, we have to determine which among the many listed factors should be assigned.

M. We classify the factors into the following three rankings by estimation:

1. Factors that have a relatively strong relationship with the target characteristic. (A rank—indicated by ◎)
2. Factors that have a strong relationship with the target characteristic. (B rank—indicated by O)
3. Factors that do not have any relationship with the target characteristic. (C rank—left unmarked)

These three rankings are marked by a double circle, a circle or no symbol. **At minimum, all factors with 1 and 2 rankings should be assigned to the orthogonal array.**

Q. I do not think the proper procedure is to anticipate the relationships between factors and characteristic values and then classify (or rank) factors based on anticipated differences in their effects before the experiment. We perform the experiment in order to confirm such effects because we cannot anticipate them. What do you think about this point?

M. I understand your question. We cannot test all of the factors actually listed, even if we use orthogonal arrays. **But the objective here is to prevent overlooking any factors that have great importance.**

Everyone must participate in factor listing so that factor importance ranking is done subjectively. Although we may still make some mistakes, ranking estimation will be more accurate if it is done by many people rather than only by the person in charge.

Q. Even if everyone agrees that a factor is effective, people may have different opinions about the magnitude of the effect. What should we do in this case?

M. Because we are simply guessing at this stage, people may well have different opinions and may not agree on the effectiveness of factors on the characteristic values. In such a case, we need not rigorously rank the factors. We assign the factors to orthogonal arrays to confirm their effect.

5-1-4. Determination of the Number of Levels

M. After factors are ranked, we determine their levels. The key points in level setting are as follows:

1. The number of levels should be either two, three, or four. Use of three levels is particularly recommended.
2. Set the range of levels widely.

Q. A larger number of levels makes it possible to more accurately describe the relationship between the output and the factor when we draw the factorial effects chart. To avoid making mistakes, why shouldn't we attempt eight or nine levels rather than three levels?

M. That is not a good idea. If you take many levels in the orthogonal array, main-effect columns will be sacrificed. Using a larger number of levels reduces the number of columns available for factors that are effective for improvement. It is better to be limit the number of levels to four for the two-level system and to three for the three-level system.

Q. We cannot describe the factorial effect precisely if the number of levels is small. Six levels gives a more precise description of factorial effects than three levels, as shown in *Figure 5-1*.

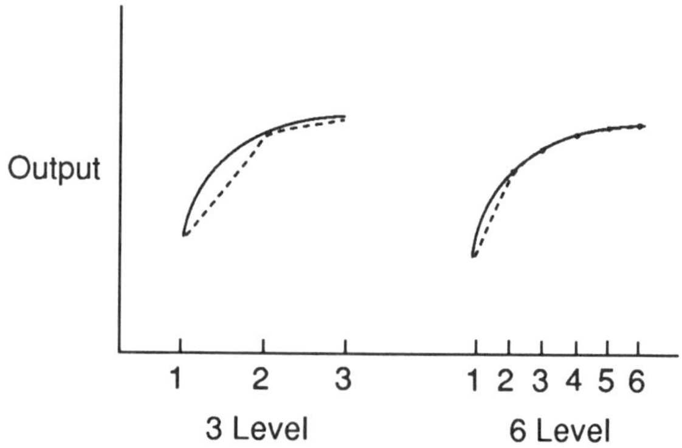

Figure 5-1. Factor Effects with 3 and 6 Levels

M. If we are only concerned about one factor and the determination of levels, six levels would be more accurate. But when we conduct orthogonal array experiments what we are concerned with is finding which factors are effective. Once we find effective factors, we can perform additional experiments on those factors to determine their suitable levels. Therefore, in experiments using orthogonal arrays, we can reduce the number of levels as long as we do not overlook factorial effects. Usually, we set the second level at the center of the target level area and the first and third levels at areas that will give nonfunctional samples. This is shown in *Figure 5-2*. **Nonfunctional samples** are those that fail to provide the objective function of a product, e.g., a motor that does not rotate or a polymerization experiment that failed to create a polymer.

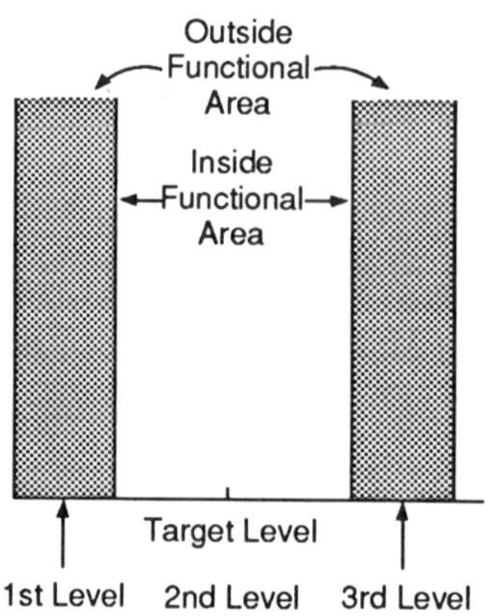

Figure 5-2: Functional Area and Level Setting

Q. I think it is wasteful to produce nonfunctional samples. They are completely different from the target product. We should search for the limits of the levels that definitely give functional samples and restrict our level setting within those limits.

M. That is the wrong mode of thinking. Yet it is one toward which most engineers frequently gravitate. The selection of the optimum level within the functional area becomes possible only when we discover the nonfunctional areas on both sides of the second level. It is quite all right if half of the samples are nonfunctional. The selection of the optimum condition is possible only from **an experimental result that has both functional and nonfunctional samples.**

Once we obtain such results, we can restrict the first and third levels to be inside the initial first and third levels, thus narrowing the limits of the acceptable levels. This is the reason that the use of **three levels is best** in the experimental design method. It is not necessary to have more than three levels.

Q. Don't we conduct experiments to accurately describe the factorial effects?

M. That is absolutely wrong. The purpose here is to confirm the influence of the factor on the output (target characteristics) by setting three levels widely. We can then determine whether a factor that has a significant influence on the output is desirable for achieving the objective. We do not need to set levels only in order to obtain a precise description of the factorial effects chart.

Q. That means that we can set the number of experimental levels at three. The range of the levels should be quite wide, shouldn't it?

M. Yes, this is right. **It is quite acceptable if more than half of the samples turn out to be nonfunctional.**

Q. If we have many nonfunctional samples, we can set the levels quite freely.

M. The above argument applies only to factors that generate continuous data. Because certain types of a catalyst or machine give discontinuous data, it is possible to have as many levels as the number of different classes or types. But even in this case, the number of levels can be reduced to three by grouping or classifying the types of machines.

Q. Because the factorial effects will be expressed as a linear line if there are only two levels, we cannot apply that to factors whose effect is expected to be nonlinear, such as temperature.

M. Usually, two levels are used for a factor whose effect is expected to increase monotonically with the level. Two levels can be set at the right and left side of the target level. This is shown in *Figure 5-3*.

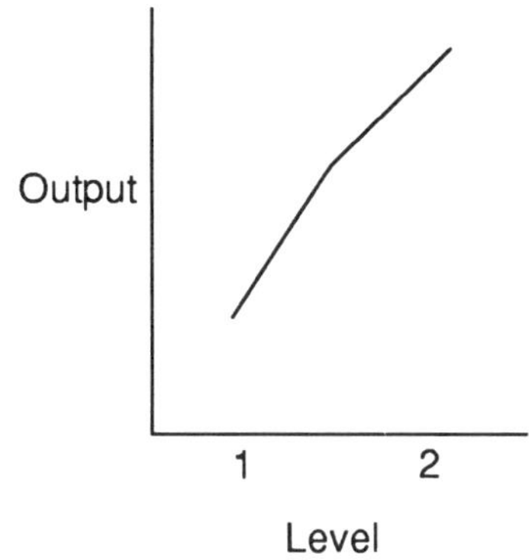

Figure 5-3: Two Level Setting for Monotonically Increasing (Decreasing) Factors

Q. What should we do if we know in advance that certain levels are not suitable for the target characteristics, e.g., a temperature that does not cause polymerization or a part size that does not fit the body?

M. A catalyst whose yield is extremely low is an example of an obviously unsuitable level. **Levels must be set within an area that is untried but which still has the possibility of achieving the goal.** An experiment is useless if it only reconfirms what doesn't work or what is wrong.

Q. Aside from the problem of quality, how do we relate level setting to considerations of cost and productivity?

M. That is a very good question. The competitiveness of a product is determined by its cost rather than its quality. Therefore, we need to select the following kinds of levels:

- Least expensive parts and process
- Least expensive processing and assembling methods
- Methods that allow for fast production and assembly in a short time
- Methods with a small number of processing steps
- Parts, processes, and systems which produce light, small, short, and thin products
- Levels that allow for a reduction in energy consumption, or number of operators, and which lead to automation
- Levels that enable flexible reaction to multi-product production, e.g., a process that reduces the die setting time

It is best to carry out research and development by first achieving the cost target (product cost plus interest, etc.) and then achieving the quality target. **If we start with the quality target, as we often do, product cost is determined afterward, and this causes a loss in competitiveness.** We find ourselves in trouble when we experience the strange phenomenon that the more research we do, the more sophisticated our parts become, and the more expensive our product becomes.

5-1-5. Level Setting

Q. Say we want to test two temperature levels, 200°C and 250°C, and it is difficult to obtain the exact temperature. In particular, longer convergence time is required in the case of large-scale equipment, causing a longer time for setting the equipment to the exact level.

M. Such a thing can happen in the actual experiment setting. In particular, continuous variables, such as temperature and density, cause this kind of problem. We can approach this problem in the following way:

1. If the experiment is done with a production machine, accept the difference if it is within the limit of daily variability. (For example, if the actual performance of the production machine is ±10°C, we start the experiment once we reach 190° through 210°C.)
2. Accept it as long as it does not interfere with the adjacent level. (In a system where the interval between the levels is small or where it is difficult to control the level, we can set the levels by simply avoiding the interference of levels. For example, the levels can be set as 175° to 225°C and 225° to 275°C.) However, it is necessary to expand the interval of levels or to improve level control capability.
3. Indicate the levels as intervals. For example, 180° to 220°C and 230° to 270°C.

In any case, the actual temperature of the experiment must be recorded as a supplementary measurement.

5-1-6. Middle Level and Level Intervals

Q. Would you explain the method for determining the actual levels?

M. Levels are usually set with equal intervals, except in certain cases where the interval is taken equiproportionally or logarithmically. These cases are summarized in *Table 5-3*.

Table 5-3: Level Interval

No.	interval	examples of levels	examples of factors
1	equal distance	20, 30, 40	temperature, pressure
2	equi-proportional	5, 10, 20	density, additive
3	equi-multiplicational	1 : 3 : 9	electric resistance, time
4	equi-exponential	10^3, 10^4, 10^5	hydrogen-ion density

We can set two end levels at the upper and lower limits of the factor, and the middle level at the target value. However, the best middle level does not necessarily coincide with the midpoint between the two end levels. The middle level must give the output level that is the average of the two output levels of the two end levels. We set the middle level by checking whether the factorial effect is increasing or decreasing exponentially or linearly. If we expect the maximum (or minimum) output level between the two end levels, the middle level is set so as to correspond to the turning point. For example, the middle level of ion density when the two end levels are 10^3 and 10^5 is 5×10^4, but 10^4 is selected based on pH. The levels should be set so that their output levels are equal.

5-1-7. Level Setting for Factors That May Interact

Q. How do we determine levels when we expect **interactions** between factors, as in the case of additives used in chemical or metallic reactions?

M. **We have to set the levels of two interacting factors so that the interaction is eliminated.**

For example, the levels of the temperature and a catalyst must be balanced. They must be as shown in *Table 5-5* rather than as shown in *Table 5-4*. The example shown in *Table 5-5* is called a **sliding level** because the middle level is shifted. Furthermore, there is a method for **compressing the level interval** in addition to the use of sliding levels. *Table 5-6* and *Table 5-7* show such an example. In cases such as that of a plasticizer, a method called the **constant ratio method** is used. Here the quantity of additive is different for different kinds of plasticizer. This example is shown in *Table 5-8*.

Table 5-4: Level Setting without Considering Interaction

A (°C)	B (g)	B_1 3	B_2 4	B_3 5
A_1	50	3	4	5
A_2	60	3	4	5
A_3	70	3	4	5

Table 5-5: Level Setting Considering Interaction (Sliding Levels)

A (°C)	B (g)	B_1 Small	B_2 Medium	B_3 Large
A_1	50	4	5	6
A_2	60	3	4	5
A_3	70	2	3	4

Table 5-6: Level Setting Without Considering Interaction

A (°C)	B (g)	B_1 25	B_2 40	B_3 55
A_1	200	25	40	55
A_2	250	25	40	55
A_3	300	25	40	55

Table 5-7: Level Setting Considering Interaction (Compressing Level Intervals)

A (°C)	B (g)	B_1 Small	B_2 Medium	B_3 Large
A_1	200	40	60	80
A_2	250	25	40	55
A_3	300	10	20	30

Table 5-8: Level Setting According to Constant Ratio

A (°C)	B (g)	B_1 Small	B_2 Medium	B_3 Large
A_1	P series	B_2 x 0.5	10	B_2 x 1.5
A_2	C series	B_2 x 0.5	25	B_2 x 1.5
A_3	H series	B_2 x 0.5	5	B_2 x 1.5

Q. How do we find the interacting factors that require this kind of level setting?

M. A chemical experiment will not be effective if the highest yield cannot be obtained from a level within the level limit. For that reason, we change the catalyst based on our expectation of the yield saturation point (maximum point) for each temperature. By doing this, **we eliminate an interaction** and also limit the levels to the appropriate level interval that can provide necessary objective information.

Q. What do we do if we cannot anticipate possible interactions between factors at all?

M. I think it is possible to anticipate possible interactions between factors in every case. We can obtain quite a lot of information from past experience. But, if we cannot anticipate interactions, we can take any of the following approaches:

1. Assign the interaction between two factors.
2. Use orthogonal arrays L_{12}, L_{18}, or L_{36} that do not generate interaction information
3. Ignore interactions when you assign factors.

I recommend using the first approach if there are unused columns. If everyone agrees that the anticipated interaction is small, the third approach should be used. The second approach is recommended if the only information required is that from main effects.

Table 5-9 lists the levels used in a case in which the quantity of polymer is altered for a different quantity of main polymer (C) in addition to the temperature difference. Twenty-seven combinations of levels exist, but if they are assigned to the first, second, and third rows of an L_{27}, the number of combinations will be nine.

Table 5-9: Level Setting of Catalyst Quantity (*B*) Against Temperature (*A*) and Polymer Quantity (*C*)

A		*C*	*B* (g)	B_1 Small	B_2 Medium	B_3 Large
		C_1	20	3	4	5
A_1	50	C_2	25	4	5	6
		C_3	30	5	6	7
		C_1	20	2	3	4
A_2	60	C_2	25	3	4	5
		C_3	30	4	5	6
		C_1	20	1	2	3
A_3	70	C_2	25	2	3	4
		C_3	30	3	4	5

Q. Can we summarize the purpose of level setting in dealing with interactions as follows? We should:

1. Eliminate interactions beforehand.
2. Eliminate combinations that do not exist in reality.

M. Yes, indeed. The ability to anticipate interactions is a part of technological know-how. It is effective to expect interactions among factors in experiments with chemical and metallic compounds and additives.

Q. When we do an orthogonal array experiment, we often obtain a nonfunctional sample (e.g., unpolymerized synthetic reaction) from experiment No. 1. The factor levels taken in experiment No.1 are usually the weakest levels, as shown in *Table 5–10*.

Table 5-10: Factors and Levels

Factor \ Level	1	2	3
Temperature (°C)	30	40	50
Rotation (rpm)	100	200	300
Quantity of polymer (%)	20	30	40

M. In such a case, we can take either of the following approaches:

1. Randomize the level to avoid a particular combination of levels.
2. Revise the levels by considering interactions.

I think the second approach is better.

Q. Is it wise to assign the levels of the current (standard) condition to exeriment No. 1?

M. If we are attempting to improve the current condition, we sometimes assign the levels of the current condition to experiment No. 1, if we are using two levels, and to experiment No. 2 if we are using three levels.

5-2. Assignment of Factors to the Orthogonal Array

Q. What is the key point in using an orthogonal array effectively?

M. The key point is to separate the use of the inner orthogonal array and the outer condition (outer orthogonal array) as follows:

Inner orthogonal array

1. Control factors are assigned to all columns (do not assign interactions or error to the columns).
2. Avoid multi-level assignment as much as possible (it is better to limit the number of levels to 4 in the case of a 2-level orthogonal array and to 3 in the case of a 3-level orthogonal array).

Outer condition

In order to determine the optimum condition based on the results of an experiment, it is necessary to maintain stable function in any environment. This is the reason that noise is added to the standard and worst conditions. The levels corresponding to these noise factors must be assigned outside of the inner orthogonal array (i.e., we must separate control factors and noise conditions).

5-2-1. Control-Factor Assignment to All Columns of the Inner Orthogonal Array

Q. I think we must assign main effects and interactions to the columns of the orthogonal array, and treat the remaining columns as error columns. **We cannot obtain error variance for testing if we do not have an error**

column. Reduction in the number of columns used for main effects makes it possible to evaluate error in the experiment and increases the reliability of the information obtained from the experiment.

M. There are statisticians who raise questions about assigning main effects to all columns of the orthogonal array. But, engineers have to think about the objective of their experiments.

Q. That is an easy problem. The objective is to create a new, inexpensive, and stable product or to come up with the optimum condition in a short period of time.

M. I think so, too. And this means that engineers perform the experiments to discover control factors with a strong enough influence on the output to achieve the target value. Even if the error column contains information about variability, it does not have information that will lead to improvement. **The engineer does not conduct the experiments in order to evaluate the error.** I do not feel the need for conducting an analysis of variance or statistical testing if it is just a mathematical expression of the impact of the factor on the output. If error variance is needed for the testing, a new error variance can be created by adding the small variation of the main effects. The error variance that is calculated from the small variation of main effects will be larger than the error variance obtained from the error column assigned under the expectation that the variability will be picked up by accident. Therefore, a test based on error variance made up of main effects is not reliable.

Q. It might be true that it improves the accuracy of testing, but I still think we should have the error column. In fact, every textbook available has an error column assigned to the orthogonal array.

M. I admit that there are many traditional experimental method textbooks that use the error column. I think the reason for this is because it is necessary for mathematical statistical analysis to have main effect, interaction, and error columns. However, because it is engineers who are applying the experimental method, it should be clear to them how important it is to assign the main effect columns in industrial experiments.

We can expect a large improvement effect only from factors that have a strong influence on the output. This information can be obtained only from the main effect columns. Therefore, the greater the number of main effects assigned to the columns, the higher the probability of achieving the engineering objective.

Conversely, inclusion of interaction and error columns means the loss of improvement information (i.e., fewer main effects).

5-2-2. Treatment of Interactions

Q. The error column may not be necessary, but I cannot accept the idea that interactions should not be assigned. A delicate condition setting (level determination) becomes simple only when we understand the interaction. For example, let us think about the interaction between factors A and B

shown in *Figure 5-4*. They say that an interaction exists between factors A and B if there is even a slight difference in the slope of B_1 and B_2. If a larger output is desirable, we can choose the better level combination, such as A_1 and B_2 or A_2 and B_2.

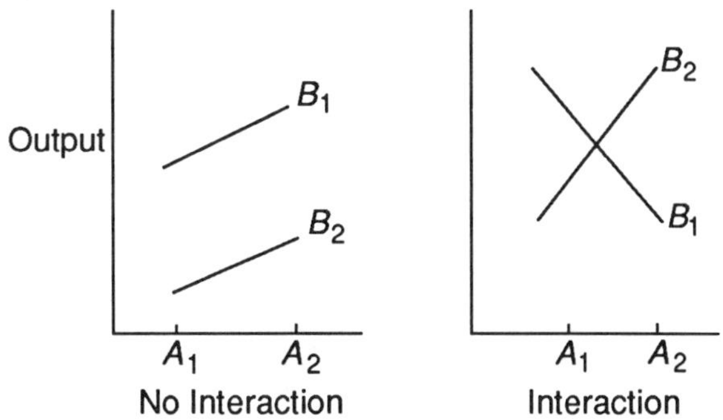

Figure 5-4: Existence and Non-existence of $A \times B$ Interaction

The ideal engineer is the one who can recognize these details.

M. Although the interaction columns will give better information than the error columns, we do not need to assign a column for the interaction, except in the following two cases:

- When we cannot anticipate the interaction between factors, and it is likely that its unknown effect will cause a significant loss if it remains unassigned
- When there are unused columns.

However, there are no cases where we cannot anticipate anything, and even if there is an unused column, we should use that for another control factor.

Figure 5-4 shows that the preferable level of B is different for a different level of A. If we accept these levels, they may cause output instability. Therefore, **the interaction between A and B is a noise component.** A **stable system** will be obtained **if condition setting (level determination) is done using only the main effect components.**

Q. But interactions always exist, and assignment of factors without thinking about this will cause confounding between the main effects and interactions, making it difficult to evaluate what is effective.

M. It is possible to separate main effects and interactions using either of the following methods: (1) assign the main effect to the points on the linear graph assuming that interactions between three factors does not exist; (2) use a special orthogonal array where interactions confound equally with every level of the main effects. The first method uses the columns corresponding to the points on the linear graph in orthogonal arrays L_8, L_{16}, L_{32}, L_9, and L_{27}. For the second method L_{12}, L_{18}, and L_{36} orthogonal arrays are used.

We can obtain the factorial effects shown as (1), (2), and (3) in *Figure 5-5* when we assign the main effect to the edge of a linear graph that contains interactions. In this case, among (1), (2), and (3), does level A_2 cause a problem if the larger output is better?

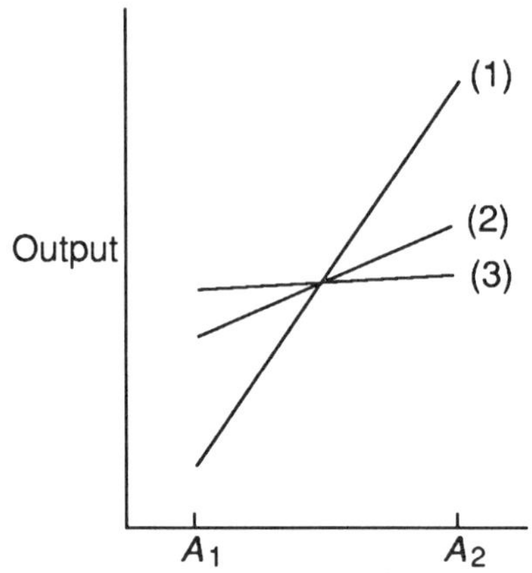

Figure 5-5: Effect of Factors Assigned to the Line of the Linear graph

Q. Assuming that the interaction of other factors is confounded with the factorial effects of A, (1) can be trusted because it has a steep slope, (3) is unreliable, and (2) is between these two. The interactions also seems to be large and the main effect small, in the case of (1), and vice-versa in the case of (3).

M. The large interaction means that the assigned main effect is small. Because any level can be taken if a factor has a small main effect, there will be no loss in choosing any level. If the main effect is larger than the interaction, the appropriate level of factors can be obtained by selecting a better level. **When the size of the main effect is the same as or smaller than the interaction, the factor has no improvement effect.** So, any level can be selected. Therefore, the choice of A_2 will not cause a loss in any of the above cases.

Q. That is clever. Let me summarize,

Case (1): Select A_2. Although the interaction may be large compared with the main effect, the choice of A_2 does not cause a loss, because any level can be chosen if the main effect is small.

Case (2): Either level will do. The weak main effect might have been the result of a smaller main effect compared to the interaction. In this case, the main effect is smaller than the interaction, which is a noise component. Therefore, either level can be selected because the improvement effect is expected to be insignificant.

Case (3): A_2 is selected for the same reason as in Case (1).

M. That is exactly right. This is not a method but a concept (philosophy). In other words, when the main effect alone is assigned and when the interaction is made to confound with the main effect, it turns out that the interaction is not significant if the estimate of the optimum condition coincides with the value obtained from the confirmatory trials. **If the estimate and confirmation test do not coincide, it proves that the interaction exists.** If they do coincide, we can expect good reproducibility even in large-scale (downstream) tests, and production is expected to be successful because there are no strong interactions. If they do not coincide, an additional experiment with different factors is required before undertaking the up-scaled experiment.

Q. We can then say that the assignment of the main effect to all columns of the orthogonal array is a good tool for judging how smoothly the development theme will be realized in mass production, can't we?

M. Yes, we can. The engineer should use the method that allows him to understand mass production from a global point of view. It is a shame that some engineers overlook an important point because they pay too much attention to small details.

5-2-3. Orthogonal Arrays and Multi-Factor (Full Factorial) Layouts

Q. Which is better, the orthogonal array or the multi-factor layout?

M. The multi-factor layout is used to obtain the cause-effect relationship (response surface) because it can give information on all interactions. If you want to describe the phenomenon precisely, you have to study the interactions.

However, **obtaining cause-effect relationships is not design.**

Q. A correct cause-effect relationship cannot be obtained if we assign main effects to all columns of the orthogonal array because the interactions confound with the main effects.

M. Obtaining cause-effect relationships from a multi-factor layout is not design. In developing a design, we determine the optimum level combination that satisfies the following conditions:

1. Variability is small and the output coincides with the target value.
2. The small variability still holds under large-scale (down-stream) conditions.
3. Low-cost products or low-cost processes can be developed in a short period of time.

In order to fulfill the **first requirement**, the optimum condition should be determined using only main effects (excluding interactions). Even if interactions are used to adjust the output to the target value, a stable condition with small variability will not be obtained.

As to the **second requirement**, the optimum condition can be considered to hold under downstream conditions if the estimate of the optimum condition coincides with the confirmatory trial results.

It does not make sense to do an experiment with a multi-factor layout, which gives information on all interactions, because interaction information is completely useless. **An orthogonal array in which only main effects are assigned to columns meets the objectives of the experimental design.**

Q. An orthogonal array experiment is also called the fractional factorial experiment. This means that it only tests particular combinations of the multi-factor layout.

M. It is true that the experiments picked up by the orthogonal array are a portion of the experiments covered in a multi-factor layout. However, for design purposes, the orthogonal array provides more effective information than does the multi-factor layout. **The belief that the multi-factor layout, which explains the cause-effect relationships based on a large number of experiments, is the best method is wrong.**

Q. But we need interactions to adjust the output. The multi-factor layout can correspond to any situation because it provides interaction information.

M. Adjustment is the lowest type of design technology. Some even exclude adjustment from the design. Adjustment is the final operation taken in the design. The number of factors tested is limited in the multi-factor layout because the number of experiments becomes prohibitively large if all level combinations are tested to check for interactions. **Because a multi-factor layout covers only a small number of factors, it could be called the partial experiment method if we take the viewpoint of design engineers.**

Q. This is an example of "one cannot see the forest for the trees", isn't it? I think it is necessary to evaluate characteristics using orthogonal arrays without being constrained by the old way of thinking.

5-2-4. The Outer Condition

We want to produce a stable product under a stable process condition which will not be affected by changes in the natural environment, usage condition, or deterioration in parts quality. This is achieved by determining the optimum condition through many experiments done under various circumstances. Conditions that induce noise in the test samples, such as the environment test and the accelerated test, are kept separate from control factors, which are assigned to the inner orthogonal array.

Noise, such as the impact of the environment and material deterioration, is classified on the basis of its source. Outer-level settings differ according to:

- Chemical change
- Mechanical change
- Electrical change
- Biological change.

There are two ways of forming the outer condition:

1. The method in which **noise is added by changing the level settings of factors slightly.** These are assigned to the inner orthogonal array. This method is used when the outer noise (e.g., changes in the environment) is considered to be the result of level changes of control factors assigned to the inner orthogonal array (inner noise). In the case of an electrical circuit, outer noises—such as temperature and humidity—cause changes, and material deterioration can be considered to be the internal change of resistance.

2. The method in which **noise is added under stressed test conditions.** Usually, the standard condition and about two other stressed (worst) conditions are used.

Q. I have heard that an American company uses 17 evaluation items. That makes it possible to obtain a reliable optimum condition.

M. But it requires too much evaluation time and effort. I recommend first determining the optimum condition based on the standard condition and two or three important evaluation items, and then checking all the evaluation items.

First, the Signal-to-Noise (S/N) ratio is calculated from the data of the worst condition and the standard condition for each experiment given by the orthogonal array. If an appropriate evaluation condition is not yet determined, the outer condition is determined by compounding the noise factors.

Q. It is important to establish the test condition suitable for the experiment, isn't it?

M. The test condition belongs to the realm of technological know-how. It is necessary for us to avoid situations in which too much effort is spent on the outer evaluation test, which results in a reduced inner orthogonal array.

Q. **The true objective of the experiment is to obtain a reliable optimum condition.** Too large an outer condition makes it necessary to expend too much effort in evaluating samples. Engineers should realize that **their objective is not sample evaluation.**

M. Developing an effective experiment involves consideration of the efficiency of running our experiments and collecting the data. Here we should be aware of the issues of replication, the order of experiments, the characteristics used for evaluation and the type of data collected.

5-3. Experimental Data Collection

5-3-1. The Number of Experimental Replications

Q. How many times should we repeat an experiment for each experimental run defined by the orthogonal array?

M. When we use an orthogonal array, the number of occurrences of the same level is given by the following:

Number of occurrences of a level = (number of replications of a level) × (number of repetitions)

Figure 5-6 shows the factorial effects of each level occurrence frequency. What frequency of level appearance is necessary in order for us to trust the result?

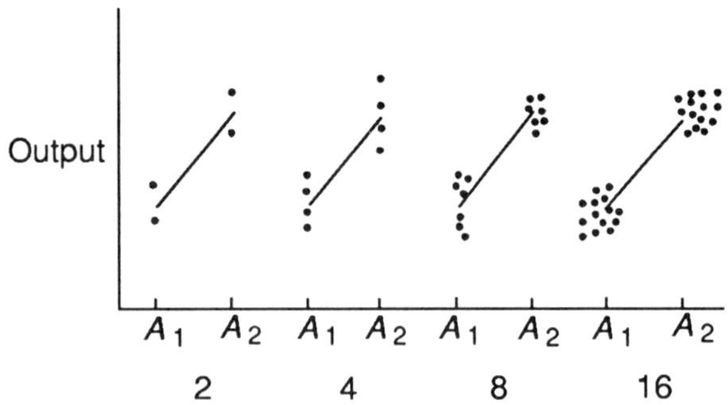

Figure 5-6: Level Occurrence Frequency and Factorial Effects.

Q. If it is more than four, we may trust the factorial effects.

M. *Table 5-11* and *Table 5-12* give the level occurrence frequency in each orthogonal array. As long as we use the orthogonal array by following the standard procedure, the number of level replications given in the orthogonal array will be sufficient.

Table 5-13 shows examples of experiments with 16 samples. Which do you think is the best?

Table 5-11: Level Occurrence Frequency in the 2-Level System

Number of Levels	L_4	L_8	L_{12}	L_{16}	L_{32}
2	2	4	6	8	16
4	—	2	—	4	8
8	—	—	—	2	4

Table 5-12: Level Occurrence Frequency in the 3-Level System

Number of Levels	L_9	L_{18}	L_{27}	L_{36}
3	3	6	9	12
6	—	3	—	—
9	—	—	3	—

Table 5-13: Experimental Situations Giving 16 Samples and the Corresponding Orthogonal Arrays

Case	Orthogonal Array	No. of Level Replications	No. of Repetitions	Level Occurrence Frequency	No. of Assigned Factors	No. of Experiments
1	L_4	2	4	8	3	16
2	L_8	4	2	8	7	16
3	L_{16}	8	1	8	15	16

Q. I think the first one is best because a larger number of repeated experiments gives a more accurate evaluation of the error component.

M. That is wrong. In this case, the third one is the best. It gives more information because the number of assigned factors is largest. It is thus possible to achieve the experimental objective with the same amount of effort.

Q. I used to think that a large number of repeated experiments increases the accuracy of the experiment and makes it possible to evaluate the experimental error.

M. That belief is wrong for two reasons. First of all, an accurate evaluation of experimental error has nothing to do with the selection of better factor levels. Secondly, **it is more important to reduce the error component by assigning more control factors than to evaluate the error component.** Finally, replication error is the most meaningless kind of error. Instead of estimating that kind of error, **it is far better to assign about two worse outer conditions and do the experiment in order to obtain the appropriate optimum condition.** After doing such an experiment, the data are analyzed to obtain the optimum condition by calculating the S/N ratio from data under the standard and worse conditions.

5-3-2. Determining of the Order of Experiments

Q. How do we determine the order of experiments? Can we arrange the order to make conducting the experiment easier?

M. *Table 5-14* shows causes of difficult experiments. If it requires a lot of effort (load) and time, we can determine the order of experiments that will make the entire experiment easy. Traditionally, we randomize the order of experiments. When it is difficult to change levels (*Table 5-15* summarizes the degree of difficulty) we should perform all of the experiments with the most difficult levels first, and then move on to experiments with less difficult levels. This experimental method is called the **split-unit design.** When doing analysis of variance (ANOVA), each factor is assigned to the linear graph according to *Table 5-16*.

Table 5-14: Causes of Difficult Experiments

No.	Reason	Constraint on Experiment
1	Takes too much time	Too many parts, disassembling difficulty, limited work space, immense production equipment
2	Too much cost	Excessive usage of raw material, recycling impossibility, high price
3	Too much effort	Equipment setting using manpower, limited number of workers

Table 5-15: Level Change and Assignment Difficulty

Factor Name	Level Change	Assignment Symbol
Primary factor	Quite difficult	○ ◎
Secondary factor	Relatively difficult	⊙
Tertiary factor	Easy	●

Table 5-16: Group Indicator of Linear Graph and Difficulty Assignment Symbol

L_{32} (2^{31})		L_{64} (2^{63})		Other Arrays	
Symbol	Group	Symbol	Group	Symbol	Group
○	1,2	○	1,2,3	○	1
◎	3	◎	4	◎	2
⊙	4	⊙	5	⊙	3
●	5	●	6	●	4

Q. How do we randomize the order of an experiment?

M. We prepare a bag containing cards with experiment numbers written on them. We then determine the order of experiments by taking one card after another out of the bag. There are other methods, too.

Q. How is assignment and data analysis done when we do the experiment based on the split-unit design?

M. There is a split-unit design in which the data are analyzed after assigning levels based on the degree of difficulty in changing levels. However, we can determine the experimental order so as to make it easy after assignment of levels following the normal method. We can then determine the optimum condition based on the factorial effects chart, and perform the confirmatory trials. Regardless whether we use analysis of variance based on the split-unit design, the optimum condition of the levels can be determined only from the factorial effects chart.

Q. But randomization is one of Fisher's Three Basic Rules. I think randomization is one of the required conditions in experimental design.

M. Randomization of the experiment is necessary to evaluate the bias of the error component. But it is virtually impossible to do randomization with a large-scale experiment in a factory or in a laboratory. If we treat randomization as an indispensable step, then any experiment will become impractical. It is more important **to assign factors with effects that are far larger than the error component** (which may cause the bias) than to make a mistake in selecting optimum levels by being overly worried about biased error allocated to particular columns.

5-3-3. Determining Evaluation Characteristics

Q. How do we determine the evaluation characteristics used at the design stage?

M. The test method and the evaluation characteristics can be classified as follows:

Test method

- Alternative (accelerated) test
- Real (quality evaluation item, shipment judgment) test

Evaluation characteristics

- Alternative (quantitative) characteristics
- Real (quality evaluation item, shipment judgment) characteristics.

At the design stage, the accelerated test and alternate characteristics are used for the evaluation.

Q. The most important evaluation items in the factory are the quality evaluation items used by the inspection department at the time of shipment. Therefore, even at the design stage, we should use the **quality evaluation items** because they are widely used and the most comprehensive.

M. **That is wrong!** The quality evaluation characteristics used before shipment are the standard for OK/NG judgment of the product. They should not be used in determining optimum levels at the design stage because their power of detection is low. **At the design stage, the evaluation must be done with an alternative (accelerated) test together with alternative continuous characteristic values**, which have high detection power, even in a limited evaluation time.

Q. What about an appearance that cannot be expressed as continuous data?

M. For characteristics that are expressed in terms of classified data, we should apply a measurement apparatus that gives continuous data. In the case of appearance, it can be measured as continuous data of color differences, glossiness, wave processing, etc. It is not difficult to quantify sensory characteristics.

Q. Because we have established methods of classified data analysis such as [0,1] and accumulation analyses, we should have no problem using classified data at the design stage.

M. We have to think about many characteristic values in the experiment, and we have to provide an analysis method that corresponds to all of these characteristic values. Even if we have accumulation analysis, I do not recommend the use of it at the design stage. Even if the output is expressed in digital numbers, it was an analog output before conversion.

Q. What you are saying is that **we should use continuous data for the design**, is that right?

M. Yes. Once the optimum condition is determined, check the result using all quality characteristics. At this point, every item relating to quality evaluation must be checked. It is acceptable for some of them to be classified data at this stage.

5-4. Data Analysis

M. In analyzing our experiment we need efficient techniques for data analysis. We need an effective way to evaluate factorial effects in order to determine an optimal design. We must be aware of how to treat nonfunctional samples, the importance of the additivity of factorial effects and the proper role of data transformation and analysis of variance in determining the optimal condition.

5-4-1. Treatment of Missing Data and Nonfunctional Samples

Q. If an experiment cannot be continued because the sample breaks down during experimentation, or if the sample is lost, we have missing data. How do we treat missing data?

M. There are three methods for treating missing data:

1. Repeat the experiment.
2. Perform the analysis by substituting the average value for the missing data.
3. Substitute the estimate obtained by the sequential approximation method for the missing data. (see Taguchi, *System of Experimental Design*, 3d ed., ASI Press, p. 840.)

The best thing is not to lose any data, but, if this happens, try the third method. It is the best of the three, and if this cannot be done, try the second method which is second best. In either case, perform the confirmatory trials after the optimum condition is determined.

Q. We often obtain nonfunctional samples from an experiment which provides:

- A motor that does not rotate
- An emulsification experiment that failed to produce emulsion
- A sensor which does not detect the signal
- Photographic film that is not photosensitive
- Seeds that do no sprout
- Parts that cannot be assembled.

Should we treat these as missing data?

M. These nonfunctional samples are proof that the experiment is successful. They contain important information for optimum level determination. **Nonfunctional data are not missing data.** They are analyzed as classified data.

Here is an example: 16 emulsion samples were made in an experiment using an L_{16} in a chemical factory. But five of them were not emulsified. The operator repeated the experiment five times for 1 condition, but none of these were emulsified, either. The operator thought that all the experiments had failed when in fact they were successful. After analyzing the data using accumulation analysis and treating the emulsion grain diameter as continuous ranked data, he succeeded in creating a milk product with the target grain diameter.

I know of many similar examples. The analysis can be done adequately even if the number of nonfunctional samples is about half of the total number of experiments. It is criminal to treat nonfunctional samples as missing data simply because the operator does not like the result.

Q. What if we find a mistake in sample condition setting after the experiment is done?

M. This is a mistake that should be distinguished from error. In such a case, either of the methods for missing data should be applied. We cannot distinguish a mistake from error in the analysis. This is an elementary level mistake of which any engineer should be ashamed.

5-4-2. Data Transformation and Additivity

Q. What is additivity?

M. Suppose there is a system with 60% reliability composed of A_1, B_1, and C_1. Suppose further that the parts were replaced with A_2, B_2, C_2 whose reliabilities are 80%, 90%, and 95%, respectively. If the reliability of each part contributes to the reliability of the system, what would be the reliability of the system as a whole after parts replacement?

Q. Because 100% is maximum, it would not exceed that, but I say it will be better than C_2.

M. There are four estimation methods:

1. The arithmetic method
2. The omega (Logit) conversion method

$$-10\log\left(\frac{1}{P}-1\right)[\text{db}]$$

3. The inverse sine transformation method

$$2\sin^{-1}\sqrt{P}$$

4. The log transformation method.

The estimation results based on these methods are shown in *Table 5-17* and *Table 5-18*.

Table 5-17: Reliability and Transformation

Part \ Level	**1** **Current**	**2** **New**
A	60	80
B	60	90
C	60	95
	(60)	

Table 5-18: Estimation Methods and Estimates

	Estimation equation	**Estimates**	**Standard**	**Gain**
1	Arithmetic method	145	60	+20 (*A*) +30 (*B*) +35 (*C*)
2	Omega (Logit) conversion method	99.8	1.761	+4.440 (*A*) +7.781 (*B*) +12.041 (*C*)
3	Inverse-sine transformation method	87.7	1.772	+0.442 (*A*) +0.726 (*B*) +0.919 (*C*)
4	Log-transformation method	190	60	$\times\frac{80}{60}$(*A*) $\times\frac{90}{60}$(*B*) $\times\frac{95}{60}$(*C*)

Q. The arithmetic method and the log transformation methods gave more than 100%, and the inverse sine transformation method gave an illogical number that is less than C_2 (95%).

M. Omega conversion is the estimation equation that gives a logically acceptable number. In this case, it can be said that **omega conversion has additivity for percentage data.**

Q. Raw data do not have additivity, do they?

M. This case concerned conversion of percentage data, but other kinds of data need appropriate transformation. *Table 5-19* summarizes the transformation methods. $\overline{y}$ is the mean of data $y_1, y_2, \ldots, y_n$, i.e.,

$$\overline{y} = \frac{1}{n}(y_1 + y_2 + \cdots + y_n)$$

Table 5-19: Important Data Transformations

	Types of Data	Transformation
1	Smaller-the-better data	$\eta = -10\log\frac{1}{n}\left(y_1^2 + y_2^2 + y_3^2 + \ldots + y_n^2\right)[\text{db}]$
2	Nominal-the-best data	• When the output is adjusted to the target value using assigned factors (parameter design): $S_m = 10\log\frac{1}{n}(y_1 + y_2 + \cdots + y_n)^2 \quad [\text{db}]$ $\eta = 10\log\left[\frac{1}{n}(SM - Ve)/Ve\right] \quad [\text{db}]$ $SM = \frac{1}{n}(y_1 + y_2 + y_3 + \cdots + y_n)^2$ $S_e = (y_1^2 + y_2^2 + \cdots + y_n^2) - SM$ $V_e = \frac{S_e}{n-1}$ • When there is an adjustment factor other than assigned factors $\eta = -10\log\frac{1}{n}\left\{(y_1 - \bar{y})^2 + (y_2 - \bar{y})^2 + \cdots + (y_n - \bar{y})^2\right\}[\text{db}]$
3	Larger-the-better data	$\eta = -10\log\frac{1}{n}\left(\frac{1}{y_1^2} + \frac{1}{y_2^2} + \cdots + \frac{1}{y_n^2}\right)[\text{db}]$
4	Data with infinity	Divide the data into intervals so as to change the data into continuous-ranked data using accumulation analysis, after counting the samples in each interval.
5	Percentage data	Perform omega (logit) conversion $\eta = -10\log\left(\frac{1}{p} - 1\right) \; [\text{db}]$
6	Classified data	Accumulate for each class. Analyze each class using accummulation analysis.

Please see Chapter 9, Section 3, for transformations of other data and characteristic values.

Q. Other textbooks use the raw data for analysis.

M. The raw data can be used for analysis because additivity will hold in a limited narrow range. However, if the data are widely dispersed, data transformation is recommended.

5-4-3. The Necessity for Analysis of Variance

Q. Analysis of variance is only discussed in the latter part of this book, isn't it? What is the role of analysis of variance?

M. I can say that the ANOVA table expresses factorial effects with numbers. The significance test, based on the ANOVA table is useful in determining which factors will be used in finding the optimum condition.

However, the optimum condition estimate, based on factors with a large factorial effect alone, gives sufficiently accurate results, as *Table 3-11* and *Table 3-13* show.

The rate of contribution of the error term is important in the ANOVA table. This is because it expresses the portion of variation that can be improved with the significant factors.

Q. I have heard that analysis of variance is not taught to engineers in the United States and in China.

M. Analysis of variance is taught to experimental design method instructors, but not to the engineers who actually apply the method. This is because **analysis of variance is not necessary for obtaining the optimum condition.** In addition, because it is almost impossible to understand even the concepts of analysis of variance in a limited lecture time, such instruction only serves to reduce the engineers' interest. It often results in their abandoning their study of the experimental design method altogether.

Q. In traditional experimental design, analysis of variance is taught first. Do you reject this method of teaching?

M. No, I do not. I only think that engineers should first study materials that give actual practical benefits and then move on to analysis of variance. This book follows that idea.

5-4-4. Determining of the Optimum Condition

M. The optimum combination of levels must give an output that coincides with the target value and reduces the impact of noise on the output.

Q. Data analysis can be done with continuous data, the S/N ratio, or accumulation analysis, etc.

M. It is possible for raw data, without being transformed, to permit selection of the optimum condition that coincides with the target value. However, in such a case, there is no guarantee that the optimum condition chosen will be stable against inner and outer noise.

Q. A better design is possible if we transform the continuous raw data into the S/N ratio, isn't it?

M. Yes, it is. The characteristic values given as Smaller-the-Better (STB), Larger-the-Better (LTB), or Nominal-the-Best (NTB) data are transformed into the S/N ratio, and the level that gives the largest S/N ratio is selected as the best level.

The following optimum condition is derived in the case of LTB data, where only assigned factors are used to reduce the variability and to achieve the target output (parameter design). Here, the general mean, *Sm*, of factors *A* and *B* and the factorial effect of the S/N ratio, η, is assumed to be what is shown in *Figure 5-7*.

1. Use η to select a level combination that gives small variability (A_2 in the case of *Figure 5-7*).
2. Next, adjust the output to the target value using a factor with a small variability effect for η and a large effect on *Sm*. (In *Figure 5-7*, change the level of *B* to achieve the output level at the target value. This makes it possible to obtain an optimum condition whose variability is small and whose output coincides with the target value.

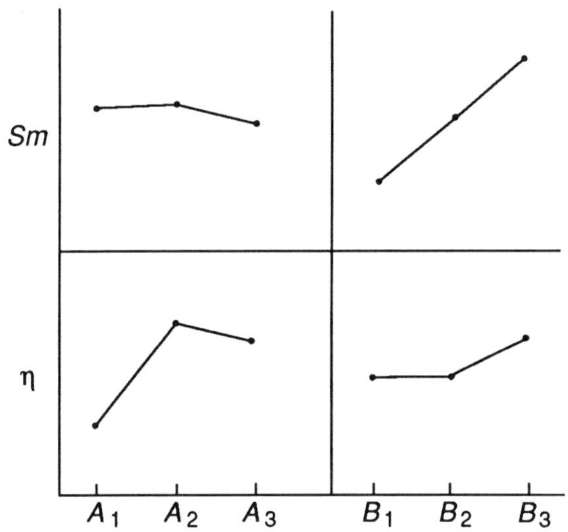

Figure 5-7: Level Determination in Parameter Design

Q. What should we do if the level of a process or a part disperses around the preset value?

M. Use a **nonlinear factor**. See *Figure 3-2* in Section 3-4-1 for details.

Q. How do we determine the optimum condition for classified data?

M. The factorial effects chart for **classified data** is shown as a bar chart from 0% to 100%. Suppose the factorial effect is assumed to be what is shown in *Figure 5-8*. Because the white portion shows the most desirable class, A_2 is the best level.

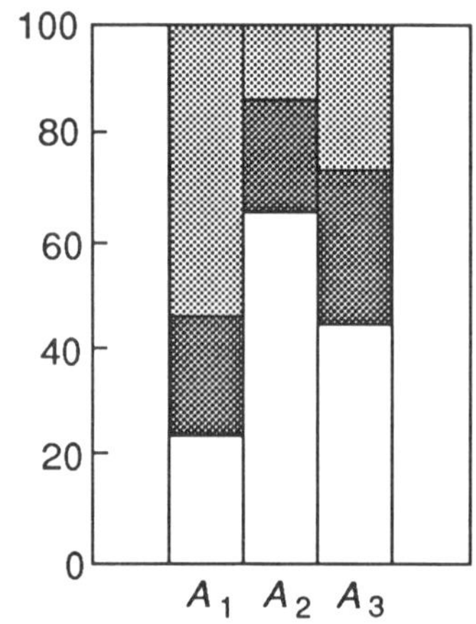

Figure 5-8: Factorial Effects for Classified Data

Q. How do we determine the optimum condition if there are **multiple characteristic values**?

M. The factorial effects for each characteristic value must be summarized in a table. The final optimum condition will be determined by examining the importance of characteristic values and factorial effects. **The trade-off determines the factor levels by considering a balance between the characteristic values.**

6

EXPERIMENTAL DESIGN METHOD EXAMPLES BASED ON THE MANUAL

This chapter introduces examples of experimental design based on the manual in Chapter 4, which covers the development/improvement and planning of experiments and data analysis. Due to space limitations, the following sections summarize each Step without going into details about checklist content.

6-1. Research on Optical-Part-Hole-Position Variability Reduction

This research relates to the drill process of an optical machine precision part. We want to drill a hole in the electronic part at a certain distance from the base position in order to automatically attach the part to the machine. If the hole position is not precisely drilled, the electronic part will not fit the machine, and it will be necessary to adjust the part by hand in a subsequent process. Furthermore, even a slight variance in hole position will be a potential cause of claims. Therefore, it is necessary to reduce hole-position variability. The absolute hole position can be set at any position by adjusting the automatic driller. Because changes in the drill process condition are believed to create burr, a decision was made to collect and check samples. Four-rank standard samples (Stepwise-grade-quality samples) were prepared.

6-1-1. Research Content

Step 1. Factor listing

The machine engineers, operators, person in charge, and manager were gathered in front of the automatic driller with sample parts containing holes to list factors relating to hole-position accuracy. The discussion was held at the location of the actual machine and samples in order to facilitate communication and understanding about the problem. In addition to this discussion within the company, the producers of the driller and the oil were interviewed beforehand to obtain their opinions about hole-position accuracy improvement. More than 50 related factors were listed. They are summarized in *Figure 6-1*.

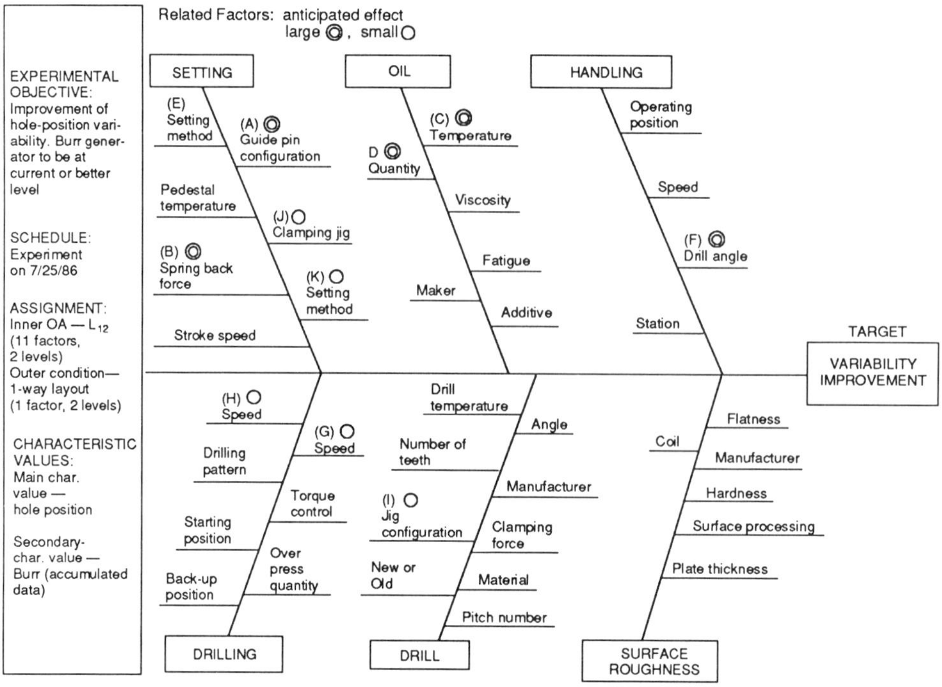

Figure 6-1: Drilling Research 7/20/86

Steps 2 – 5. Factor classification and level determination

The factors listed were classified into control, indicative, block, and noise factors. The anticipated effect of control factors was discussed. Factors thought to have a significant effect were marked with a double circle. Those thought to have a moderate effect were marked with a single circle, as shown in *Figure 6-1*.

After an examination of the levels of each factor, factor levels were determined as shown in *Table 6-1*.

Steps 6 – 9. Selection of orthogonal arrays and preparation of experiment instructions

It was decided that only one day would be spent on the experiment and that the number of experiments would be about 15. An L_{12}, L_{16}, or L_{18} orthogonal array could be used. In this case, an L_{12} was chosen because there were many factors that had two levels and the number of factors to be tested was about ten.

The following are the characteristics of an L_{12}:

1. Because interactions are equally distributed among the columns, we can use each column to evaluate the main effect free of interaction.
2. Because main effects are assigned to every column, it is possible to obtain design information efficiently and effectively.

For the outer condition, two levels of pedestal temperature, which is a noise factor affecting variability, were used. *Table 6-1* shows the experimental instructions corresponding to factor assignment in the order of the experiment. Because an L_{12} is an orthogonal array that has only main effect columns,

Table 6-1a: Assignment, Experimental Instructions, and Data

SYMBOL	*A*	*B*	*C*	*D*	*E*	*F*	*G*	*H*	*I*	*J*	*K*
Factor / Column / No.	Guide Pin 1	Spring Press Force 2	Oil Temp. 3	Oil Quantity 4	Setting Method 5	Insertion Angle 6	Speed 7	Process-ing Speed 8	Clamp 9	Support-ing Point 10	Strike Speed 11
1	1	1	1	1	1	1	1	1	1	1	1
2	1	1	1	1	1	2	2	2	2	2	2
3	1	1	2	2	2	1	1	1	2	2	2
4	1	2	1	2	2	1	2	2	1	1	2
5	1	2	2	1	2	2	1	2	1	2	1
6	1	2	2	2	1	2	2	1	2	1	1
7	2	1	2	2	1	1	2	2	1	2	1
8	2	1	2	1	2	2	2	1	1	1	2
9	2	1	1	2	2	2	1	2	2	1	1
10	2	2	2	1	1	1	1	2	2	1	2
11	2	2	1	2	1	2	1	1	1	2	2
12	2	2	1	1	2	1	2	1	2	2	1
First Level	Current	Current	Current	Current	Current	Current	Current	Current	Current	Current	Current
Second Level	Round type	Current +20%	Current −15°C	Current −50%	New method	Current −10%	Current +30%	Double the current	New method	4 point	2 step

Table 6-1b: Assignment, Experimental Instructions and Data

	Observed (Raw) Data													Transformed Data				
	Hole Position (Measured Value — Base Value)										Burr Generation			Hole Position		Burr Generation		
	Pedestal Temperature										Good O	Moderate Δ	Inferior X	Average value y	S/N	Accumulated Data I	II	III
1	−24	10	39	4	0	−23	11	22	0	−19	14	6	0	5.60	-24.937	14	20	20
	15	14	3	-19	19	21	9	-5	-3	38								
2	21	16	3	1	25	23	8	-8	2	28	19	1	0	15.35	-21.548	19	20	20
	1	26	30	32	4	6	20	33	19	17								
3	77	35	36	67	65	47	15	48	14	48	10	2	8	44.75	-27.150	10	12	20
	33	17	74	57	46	66	19	36	88	7								
4	42	32	19	27	-64	24	25	33	69	16	15	4	1	21.50	-28.495	15	19	20
	13	18	24	29	12	67	17	-7	-1	35								
5	35	30	46	29	30	0	67	32	39	20	17	3	0	32.60	-24.037	17	20	20
	9	37	45	45	58	28	45	24	24	9								
6	68	85	87	81	31	40	47	14	16	68	9	8	3	56.30	-28.041	9	17	20
	67	31	31	38	53	103	60	54	98	54								
7	-4	-13	15	21	-5	21	-18	2	34	-4	2	6	12	2.75	-24.005	2	8	20
	-11	-5	13	-14	1	13	-3	5	-25	32								
8	38	30	66	17	6	4	23	6	30	21	4	11	5	24.10	-24.024	4	15	20
	36	30	36	45	0	23	9	35	15	12								
9	76	49	75	76	-6	26	36	54	89	134	14	5	1	48.65	-30.189	14	19	20
	32	60	9	34	42	64	72	14	24	13								
10	8	-10	12	20	-2	25	14	19	6	14	18	2	0	7.55	-19.980	18	20	20
	-4	8	-7	6	-6	-3	16	3	13	19								
11	34	17	6	31	24	12	36	0	19	9	7	3	10	18..80	-21.930	7	10	20
	25	26	5	0	30	13	27	0	42	20								
12	13	-19	-18	-28	-19	-25	0	-28	-10	-26	1	7	12	-14.75	-22.438	1	8	20
	-21	5	14	-26	-10	-5	-28	-18	-17	-28								

the factors can be assigned starting at column 1 in the order decided on for the experiment.

Steps 10 – 11. Data transformation

There were no nonfunctional samples, such as samples without a hole. There were no missing data caused by sample breakdown. The hole-position data were calculated by subtracting the base value from the measured value. The degree of burr generation was classified into three ranks, i.e., good, moderate, and inferior, because no samples were drastically inferior, falling short of the standard. This is shown in the entry of observed (raw) data in *Table 6-1*.

Next, the data were transformed as follows:

1. Data transformation of the hole position

The center of the hole position can be adjusted freely using the automatic driller. In order to evaluate variability, the data are treated as Smaller-the-Better (STB) data whose target value is the average of the entire data ($\bar{y}$). The transformation follows the following formula:

$$\eta = -10\log\frac{1}{20}\left\{(y_1-\bar{y})^2+(y_2-\bar{y})^2+\cdots+(y_{20}-\bar{y})^2\right\}$$

$$= -10\log\sigma^2 \quad [\text{db}]$$

σ^2 is called the variance, and η becomes large as σ^2 becomes small. η was calculated for experiments 1 – 12 as follows:

$$\text{No.1}\quad \eta_1 = -10\log\frac{1}{20}\{(-24-5.6)^2+(10-5.6)^2+\cdots$$

$$+(38-5.6)^2\}$$

$$= -24.937 \quad [\text{db}]$$

$$(\bar{y}_1 = \frac{1}{20}\{-24+10+39+\cdots+38\} = \frac{112}{20} = 5.6)$$

$$\text{No.2}\quad \eta_2 = -10\log\frac{1}{20}\{(21-15.35)^2+(16-15.35)^2+\cdots$$

$$+(17-15.35)^2\}$$

$$= -21.548 \quad [\text{db}]$$

$$(\bar{y}_2 = \frac{1}{20}\{21+16+3+\cdots+17\} = \frac{307}{20} = 15.35)$$

$$\vdots$$

$$\text{No.12}\ \eta_{12} = -10\log\frac{1}{20}\{(13-(-14.7))^2+(-19-(-14.7))^2+\cdots$$

$$+(-28-(-14.7))^2\}$$

$$= -22.438 \quad [\text{db}]$$

$$(\bar{y}_{12} = \frac{1}{20}\{13-19-18-\cdots-28\} = \frac{-294}{20} = -14.7)$$

2. Data transformation for burr data

Twenty sample pieces were compared to the Stepwise-grade-quality samples, and the number of samples was treated as accumulated data.

Class I were considered good; Class II, good plus moderate; and Class III, good plus moderate plus inferior. The accumulated data shown in the following example were obtained for experiments No. 1 – 12.

Observed (Raw) Data				Accumulated Data		
No.	Good	Moderate	Inferior	I	II	III
1	14	6	0	14	20	20
2	19	1	0	19	20	20
12	1	7	12	1	8	20

No. 1: Class I = 14 = 14

Class II = 14 + 6 = 20

Class III = 14 + 6 + 0 = 20

No. 12: Class I = 1 = 1

Class II = 1 + 7 = 8

Class III = 1 + 7 + 12 = 20

Steps 12 – 13. Level sum, and process average

The level sum and the process average were calculated as shown below.

1. Level sum and process average of the hole position

The level sum and the process average of the hole position were calculated as follows:

Level sum of A_1 $A_1 = (-24.937) + (-21.548) + \cdots + (-28.041) = -154.21$ db

Level sum of A_2 $A_2 = (-24.005) + (-24.024) + \cdots + (-22.438) = -142.57$ db

The process averages were obtained by dividing the level sum by the number of data:

$$\text{Average of } A_1 \quad \overline{A}_1 = \frac{-154.21}{6} = -25.70 \text{ db}$$

$$\text{Average of } A_2 \quad \overline{A}_2 = \frac{-142.57}{6} = -23.76 \text{ db}$$

The level sum and the process average of other factors were calculated similarly. The following shows the case of factor K:

Level sum of K_1 $K_1 = (-24.937) + (-24.037) + \cdots + (-22.438) = -153.65$ db

Level sum of K_2 $K_2 = (-21.548) + (-27.150) + \cdots + (-21.930) = -143.13$ db

$$\text{Average of } K_1 \quad \overline{K}_1 = \frac{-153.65}{6} = -25.61 \text{ db}$$

$$\text{Average of } K_2 \quad \overline{K}_2 = \frac{-143.13}{6} = -23.85 \text{ db}$$

The results are summarized in *Table 6-2*.

Table 6-2: Level Sum and Process Average for Hole Position (db)

Factor \ Level	Level Sum 1	Level Sum 2	Process Average 1	Process Average 2
A	–154.21	–142.57	–25.70	–23.76
B	–151.85	–144.92	–25.31	–24.15
C	–149.54	–147.24	–24.92	–24.54
D	–136.96	–159.81	–22.83	–26.64
E	–140.44	–156.33	–23.41	–26.06
F	–147.01	–149.77	–24.50	–24.96
G	–148.22	–148.55	–24.70	–24.76
H	–148.52	–148.25	–24.75	–24.71
I	–147.43	–149.35	–24.57	–24.89
J	–155.57	–141.11	–25.94	–23.52
K	–153.65	–143.13	–25.61	–23.85
(*T*)	(–296.78)		(–24.73)	

Due to computer output restrictions, this table shows only two digits below zero. It is recommended that three digits below zero be used.

2. Level sum and average of the burr data

The level sum was calculated for each accumulation class:

Class I Level sum $\overline{A}_1 = 14 + 19 + 10 + \cdots + 9 = 84$

Class II Level sum $\overline{A}_2 = 20 + 20 + 20 + \cdots + 17 = 108$

Class III Level sum $\overline{A}_3 = 20 + 20 + 20 + \cdots + 20 = 120$

The process average was calculated in terms of percentage (%):

Class I Average $\overline{A}_1 = \left(\frac{84}{120}\right) \times 100 = 70(\%)$

Class II Average $\overline{A}_2 = \left(\frac{108}{120}\right) \times 100 = 90(\%)$

Class III Average $\overline{A}_3 = \left(\frac{120}{120}\right) \times 100 = 100(\%)$

Next, the level sum and the average of K_1 were calculated:

Class I Level sum $\overline{K}_1 = 14 + 17 + 9 + \cdots + 1 = 57$

Class II Level sum $\overline{K}_2 = 20 + 20 + 17 + \cdots + 8 = 92$

Class III Level sum $\overline{K}_3 = 20 + 20 + 20 + \cdots + 20 = 120$

The average of K_1 for each class was calculated by setting that of Class III at 100%:

$$\text{Class I} \quad \text{Average} \quad \overline{K}_1 = \left(\frac{57}{120}\right) \times 100 = 47.5(\%)$$

$$\text{Class II} \quad \text{Average} \quad \overline{K}_2 = \left(\frac{92}{120}\right) \times 100 = 76.67(\%)$$

$$\text{Class III} \quad \text{Average} \quad \overline{K}_3 = \left(\frac{120}{120}\right) \times 100 = 100(\%)$$

A similar calculation was done for other factors. The results are summarized in *Table 6-3*.

Table 6-3: Level Sum and Average for Burr Generation

		Level Sum			Average (%)		
Factor	Level	I	II	III	I	II	III
A	1	84	108	120	70.00	90.00	100.00
	2	46	80	120	38.33	66.67	100.00
B	1	63	94	120	52.50	78.33	100.00
	2	67	94	120	55.83	78.33	100.00
C	1	70	96	120	58.33	80.00	100.00
	2	60	92	120	50.00	76.67	100.00
D	1	73	103	120	60.83	85.83	100.00
	2	57	85	120	47.50	70.83	100.00
E	1	69	95	120	57.50	79.17	100.00
	2	61	93	120	50.83	77.50	100.00
F	1	60	87	120	50.00	72.50	100.00
	2	70	101	120	58.33	84.17	100.00
G	1	80	101	120	66.67	84.17	100.00
	2	50	87	120	41.67	72.50	100.00
H	1	45	82	120	37.50	68.33	100.00
	2	85	106	120	70.83	88.33	100.00
I	1	59	92	120	49.17	76.67	100.00
	2	71	96	120	59.17	80.00	100.00
J	1	74	110	120	61.67	91.67	100.00
	2	56	78	120	46.67	65.00	100.00
K	1	57	92	120	47.50	76.67	100.00
	2	73	96	120	60.83	80.00	100.00
T		130	188	240	54.17	78.33	100.00

Step 15. Drawing the factorial effects chart

The averages summarized in *Table 6-2* and *Table 6-3* were used to draw the factorial effects chart for hole position and burr generation. These are shown in *Figure 6-2* and *Figure 6-3*.

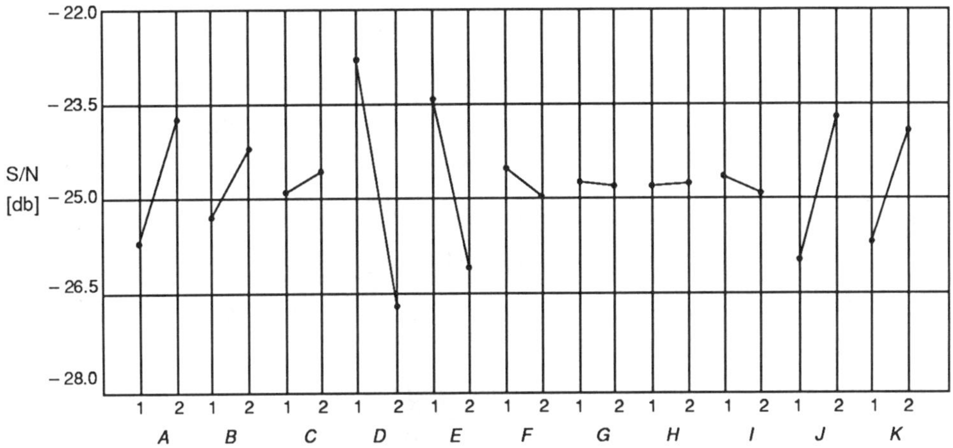

Figure 6-2: Factorial Effects — Hole Position

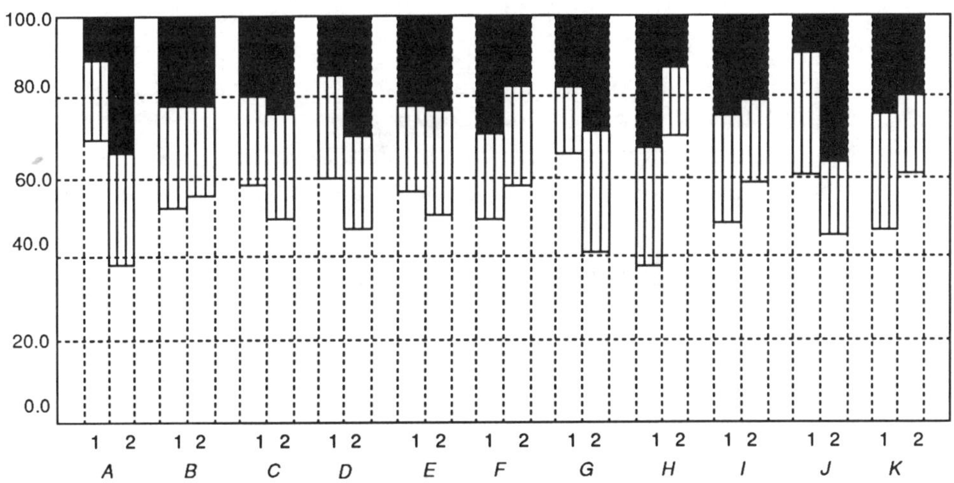

Figure 6-3: Factorial Effects — Burr Generation

Step 16. Estimation of the optimum condition

Figure 6-2 shows that the factors that are effective in reducing the variability of hole position are *A, B, D, E, J,* and *K*. Their optimum levels are A_2, B_2, D_1, E_1, J_2, and K_2. The factors that are strongly related to burr generation are *A, D, G, H,* and *J*. Their optimum levels are A_1, D_1, G_1, H_2, and J_2. The optimum levels of factors *A* and *J* are different for hole position and burr generation. Their levels were determined to be A_2 and J_2 because hole position is the important (main) characteristic. Although burr generation is an important characteristic, it is a secondary-characteristic. In order to improve the burr characteristic, levels G_1 and H_2 were selected because factors G and H do not affect hole position. Also, D_1 is the best level for both characteristics. Factors C, F, and I do not have any significant effect on either characteristic. The levels of these factors can be determined on the basis of cost and productivty. The final optimum level was determined as A_2, B_2, C_1, D_1, E_1, F_1, G_1, H_2, I_1, J_2, and K_2.

Next, estimates of the optimum condition for hole position and burr generation were derived.

1. Hole-position estimate of optimum condition

The estimate of the optimum condition for hole position is derived as:

$$\hat{\mu}(A_2B_2D_1E_1J_2K_2) = -23.761 - 24.154 - 22.827 - 23.407 - 23.518$$
$$-23.855 - 5 \times (-296.78 / 12)$$
$$= -17.865\,[\text{db}]$$

The variance for hole position, for the current condition (No.1) and for the optimum condition was derived:

$$\text{Current condition: } -24.937 = -10 \log \sigma^2_{\text{current}}$$
$$\sigma^2_{\text{current}} = 311.67$$
$$\text{Optimum condition: } -17.865 = -10 \log \sigma^2_{\text{optimum}}$$
$$\sigma^2_{\text{optimum}} = 61.16$$

The variance of the optimum condition ($\sigma^2_{\text{optimum}}$) is expected to be about one-fifth that of the current condition ($\sigma^2_{\text{current}}$).

2. Estimate of the Optimum Condition for Burr Generation

This estimate is based only on factors that are related to burr generation (large-effect factors). Estimates for Class I and II were obtained separately.

$$\text{Class I} \quad \hat{\mu}(A_2D_1G_1H_2J_2) = \overline{A}_2(38.33\%) + \overline{D}_1(60.83\%) + \overline{G}_1(66.67\%)$$
$$+\overline{H}_2(70.83\%) + \overline{J}_2(46.67\%) - 4 \times \overline{T}(54.17)$$
$$= -2.065 + 1.912 + 3.010 + 3.854 - 0.580$$
$$-4 \times 0.7255$$
$$= 3.229\,[\text{db}]$$
$$= 67.78\,(\%)$$

$$\text{Class II} \quad \hat{\mu}(A_2D_1G_1H_2J_2) = 3.010 + 7.824 + 7.256 + 8.792 + 2.688$$
$$-4 \times 5.5815$$
$$= 7.244\,[\text{db}]$$
$$= 84.13\,(\%)$$

When these are compared to the current condition (No.1), the estimate for Class II is not as good as the current condition. However, this condition was accepted and the confirmatory trial was done based on this condition, because the burr characteristic is a sub-characteristic and there were no defects among the samples.

3. Confirmatory trial

The automatic driller was set to the optimum condition and 20 samples were made. The variability of the hole position was -17.421 [db] ($\sigma^2 = 55.22$), an improvement of 7.5 [db]. Burr generation was the same as the current condition. When the optimum condition was applied to the production machine, the problem in the automatic part-assembling process was eliminated.

6-1-2. Discussion of the Study

Q. What do you think about analyzing the data in this example by setting the target value at $\overline{Z}$ and transforming the data according to the following formula?

$$\eta = -10\log\frac{1}{20}\{\Sigma(y_i - \overline{Z})^2\}$$

M. η in this formula includes the length of bias of the average of y_i from $\overline{Z}$. Since there are other adjustment factors in this equipment, it is better to use the data transformation formula that minimizes the difference between y_i and $\bar{y}$ (average value) that does not contain the bias.

The base position used in the observed (raw) data in *Table 6-2* is the target hole position. Let us compare the data for No. 7 and No. 8. The variability from the mean is about the same for both No. 7 and No. 8, i.e., –24 db, but the variability from the target value is quite different, i.e., 2.75 for No. 7 and 24.1 for No. 8. If this bias is included in the evaluation, No. 8 will be judged to be far worse than No. 7.

Q. The optimum condition improves the current condition by one-fifth in terms of variability, but how much would this improvement be if it were expressed in monetary terms?

M. That can be calculated from the Quality Loss Function.

A: Average loss from out-of-standard product: 5000 ¥/unit

Δ: Tolerance ± 70

Yearly production: 12,000 units/year

Current loss: $$L_{current} = \frac{A}{\Delta^2} \times \sigma^2 = \frac{5000}{70^2} \times 311.67 = 318.03 \text{ ¥ / unit}$$

Loss after improvement: $$L_{optimum} = \frac{A}{\Delta^2} \times \sigma^2 = \frac{5000}{70^2} \times 55.22 = 56.35 \text{ ¥ / unit}$$

The monetary improvement for one year is:

Quantity of production × ($L_{current}$ − $L_{optimum}$)
=12000 × (318.03 - 56.35)
= ¥3,140,160
≅ ¥3.14 mil/year

Q. This is convincing because the improvement is evaluated in monetary terms.

M. It used to be the case that monetary evaluation was possible only for cost improvement but not for variation improvement. In this sense, the Quality Loss Function is an epoch making device that allows us to evaluate quality improvement in terms of money. This function can also be applied to (STB) and Larger-the-Better (LTB) data.

6-2. Improvement of Picture Image Formation—Film Adhesiveness and Picture Developability

This research relates to the following picture formation method: A thermoplastic-type photosensitive resin is coated on a glass plate, and the glass plate is exposed to ultraviolet light. The unexposed portion of the resin is then removed with developer.

The thermoplasticized portion needs to be tightly bonded to the glass plate, and the part that was unexposed needs to be removed completely. In other words, the photosensitive film must satisfy two conflicting properties simultaneousely: adhesiveness and picture developability. Adhesiveness here is an LTB characteristic, and picture developability is an STB characteristic. This research intends to develop a photosensitive resin whose adhesiveness level is above 35 and whose picture developability is below 25.

6-2-1. Research Content

Step 1. Factor listing

Photosensitivity and picture developability experts and others related to the project were gathered to participate in listing factors that could improve the adhesiveness and picture developability of a new photosensitive material. The results of the discussion are summarized in *Figure 6-4*.

Step 2. Factor classification and determination of levels

The listed factors were classified into control, indicative, block, and noise factors. The effect of control factors was evaluated. Factors were marked with a double circle if their effect was considered large and with a single circle if considered moderate. The levels of assigned factors are listed in *Table 6-4* .

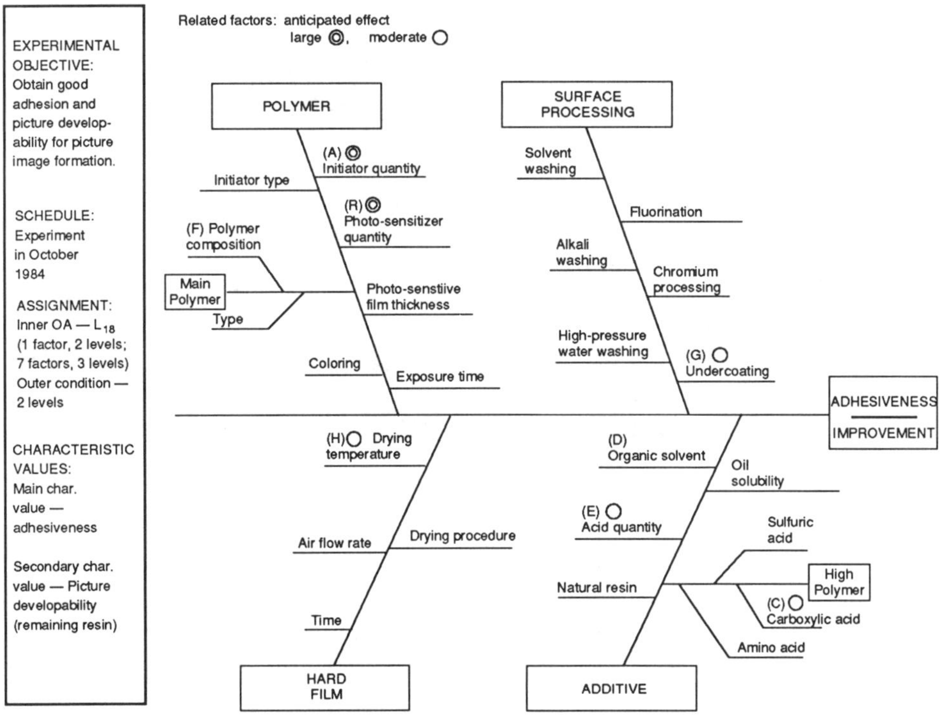

Figure 6-4: Picture Image Formation Film Research

Steps 6–9. Selection of orthogonal arrays and preparation of experiment instructions

An L_{18} was selected because the number of experiments had to be limited to about 20 and seven or eight factors with three levels were to be tested.

Characteristics of the L_{18} Array:

1. Because interactions are equally distributed among columns, we can use each column to evaluate the main effect free of interactions.
2. An L_{18} provides information on main effect columns with three levels. This is effective in developing the design.

An enforcement condition other than the standard condition was used as the outer noise condition. The enforcement condition was a 20% reduction in adhesiveness and four-fifths solution for the development.

The assignment and experiment instructions are given in *Table 6-5.*

Steps 10–11. Data transformation

None of the samples obtained from the experiment were nonfunctional; in other words, none of the photosensitive parts were peeled off during development, and the picture was developed in every sample. There were no missing data caused by sample breakdown. The raw data are shown in *Table 6-5.*

Table 6-4: Factors and Levels

Sign	Factor Name	First Level	Second Level	Third Level
A	Ignition quantity as a proportion of *B*	Small (0.08 x *B*)	Large (0.12 x *B*)	—
B	Photo-sensitizer quantity	2	3	4
C	Carboxylic acid type additive quantity	2.5	3.0	3.5
D	Organic solvent ratio	2/8	3/7	4/6
E	Oxygen quantity (%)	0.3	0.4	0.5
F	Polymer ratio	4/6	5/5	6/4
G	Undercoating quantity (mg)	0	1	2
H	Drying temperature (C°)	60	70	80

The data were transformed as follows:

1. Adhesiveness data transformation

   Because adhesiveness is given as LTB data, it transformed into the Signal-to-Noise (S/N) ratio according to the following formula:

$$\eta = -10\log\frac{1}{n}\left(\frac{1}{y_1^2}+\frac{1}{y_2^2}+\cdots+\frac{1}{y_n^2}\right) \quad [\text{db}]$$

The S/N ratio for each experiment was calculated as follows:

$$\text{No.1} \quad \eta_1 = -10\log\frac{1}{2}\left(\frac{1}{48^2}+\frac{1}{47^2}\right) = 33.532 \quad (\text{db})$$

$$\text{No.2} \quad \eta_2 = -10\log\frac{1}{2}\left(\frac{1}{39^2}+\frac{1}{38^2}\right) = 31.707 \quad (\text{db})$$

⋮

⋮

$$\text{No.18} \quad \eta_{18} = -10\log\frac{1}{2}\left(\frac{1}{36^2}+\frac{1}{24^2}\right) = 29.018 \quad (\text{db})$$

Table 6-5a: Assignment, Experimental Instruction and Data

Orthogonal Array L_{18} ($2^1 \times 3^7$)									**Experiment Instructions**							
Factor	*A*	*B*	*C*	*D*	*E*	*F*	*G*	*H*	Under-coating (*G*)	Initi-ation (*A*)	Photo-sensitizer (*B*)	Carbox-ylic Acid (*C*)	Solvent Ratio (*D*)	Oxygen Quantity (*E*)	Polymer Ratio (*F*)	Drying Temper-ature (*H*)
Column No.	1	2	3	4	5	6	7	8	7	1	2	3	4	5	6	8
1	1	1	1	1	1	1	1	1	0	0.16	2	2.5	2/8	0.3	4/6	60
2	1	1	2	2	2	2	2	2	1	0.16	2	3.0	3/7	0.4	5/5	70
3	1	1	3	3	3	3	3	3	2	0.16	2	3.5	4/6	0.5	6/4	80
4	1	2	1	1	2	2	3	3	2	0.24	3	2.5	2/8	0.4	5/5	80
5	1	2	2	2	3	3	1	1	0	0.24	3	3.0	3/7	0.5	6/4	60
6	1	2	3	3	1	1	2	2	1	0.24	3	3.5	4/6	0.3	4/6	70
7	1	3	1	2	1	3	2	3	1	0.32	4	2.5	3/7	0.3	6/4	80
8	1	3	2	3	2	1	3	1	2	0.32	4	3.0	4/6	0.4	4/6	60
9	1	3	3	1	3	2	1	2	0	0.32	4	3.5	2/8	0.5	5/5	70
10	2	1	1	3	3	2	2	1	1	0.24	2	2.5	4/6	0.5	6/4	60
11	2	1	2	1	1	3	3	2	2	0.24	2	3.0	2/8	0.3	6/4	70
12	2	1	3	2	2	1	1	3	0	0.24	2	3.5	3/7	0.4	4/6	80
13	2	2	1	2	3	1	3	2	2	0.36	3	2.5	3/7	0.5	4/6	70
14	2	2	2	3	1	2	1	3	0	0.36	3	3.0	4/6	0.3	5/5	80
15	2	2	3	1	2	3	2	1	1	0.36	3	3.5	2/8	0.4	6/4	60
16	2	3	1	3	2	3	1	2	0	0.48	4	2.5	4/6	0.4	6/4	70
17	2	3	2	1	3	1	2	3	1	0.48	4	3.0	2/8	0.5	4/6	80
18	2	3	3	2	1	2	3	1	2	0.48	4	3.5	3/7	0.3	5/5	60

Table 6-5b

Raw Data				Transformed Data	
Adhesiveness		Picture Developability (Remaining Resin)		Adhesiveness (LTB Characteristic) [db]	Picture Developability (STB Characteristic) [db]
Standard Condition	Accelerated Condition	Standard Condition	Accelerated Condition		
48	47	40	43	33.532	−32.367
39	38	22	24	31.707	−27.243
29	27	17	19	28.927	−25.119
47	45	39	41	33.249	−32.044
38	36	42	45	31.355	−32.775
29	26	46	48	28.748	−33.444
26	25	55	58	28.126	−35.044
32	31	49	51	29.963	−33.981
24	23	54	56	27.415	−34.809
39	37	29	32	31.587	−29.696
36	31	21	33	30.428	−28.837
35	35	38	40	30.881	−31.824
38	37	37	41	31.478	−31.833
37	35	50	52	31.116	−34.153
36	35	43	45	31.002	−32.871
34	33	55	58	30.498	−35.044
26	26	46	47	28.299	−33.350

2. Picture developability data transformation

The picture-developability data were obtained from the quantity of remaining resin on the unexposed part of the glass plate after development. The quantity of remaining resin is STB data. The STB data is transformed into the S/N ratio according to the following formula:

$$\eta = -10\log\frac{1}{n}(y_1^2 + y_2^2 + \cdots + y_n^2) \quad \text{[db]}$$

The S/N ratio for each experiment is calculated as follows:

$$\text{No.1} \quad \eta_1 = -10\log\frac{1}{2}(40^2 + 43^2) = -32.367 \quad \text{[db]}$$

$$\text{No.2} \quad \eta_2 = -10\log\frac{1}{2}(22^2 + 24^2) = -27.243 \quad \text{[db]}$$

⋮

⋮

$$\text{No.18} \quad \eta_{18} = -10\log\frac{1}{2}(52^2 + 55^2) = -34.570 \quad \text{[db]}$$

Transformed data are summarized on the right hand side of *Table 6-5*.

Steps 12–13. Level sum and process average

The level sum and the process average were calculated as follows.

1. Adhesiveness level sum and process average

The level sums and process averages of adhesiveness were calculated as follows.

The level sum can be obtained as the sum of transformed data corresponding to the level of columns assigned with the factor. For example, the level sums of factors A and B are as follows:

$$A_1 = 33.532 + 31.707 + \cdots + 27.415 = 272.84$$
$$A_2 = 31.587 + 30.428 + \cdots + 29.018 = 274.31$$
$$B_1 = 33.532 + 31.707 + \cdots + 30.881 = 186.88$$
$$B_2 = 33.249 + 31.355 + \cdots + 31.002 = 186..95$$
$$B_3 = 28.126 + 29.963 + \cdots + 29.018 = 173.32$$

The average was obtained by dividing the level sum by the number of data:

$$\overline{A}_1 = \frac{272.84}{9} = 30.32 \qquad \overline{B}_1 = \frac{186.88}{6} = 31.15$$

$$\overline{A}_2 = \frac{274.31}{9} = 30.48 \qquad \overline{B}_2 = \frac{186.95}{6} = 31.16$$

$$\overline{B}_3 = \frac{173.32}{6} = 28.89$$

The level sum and average of other factors were calculated similarly.

2. Picture Developability level sum and average

A procedure similar to the preceding example was used to obtain the following:

$A_1 = -286.83 \qquad \overline{A}_1 = -31.87$

$A_2 = -292.18 \qquad \overline{A}_2 = -32.46$

$B_1 = -175.09 \qquad \overline{B}_1 = -29.18$

$B_2 = -197.12 \qquad \overline{B}_2 = -32.85$

$B_3 = -206.80 \qquad \overline{B}_3 = -34.47$

These level sums and averages are shown in *Table 6-6* and *Table 6-7*.

Table 6-6: Adhesiveness—Level Sum and Average

Level / Factor	Level Sum			Average		
	1	2	3	1	2	3
A	272.84	274.31	—	30.32	30.48	—
B	186.88	186.95	173.32	31.15	31.16	28.89
C	188.29	182.87	175.99	31.38	30.48	29.33
D	183.75	182.57	180.84	30.62	30.43	30.14
E	180.79	187.30	179.06	30.13	31.22	29.84
F	182.72	184.09	180.34	30.45	30.68	30.06
G	184.62	179.47	183.06	30.77	29.91	30.51
H	186.28	180.27	180.60	31.05	30.05	30.10
(*T*)	(547.149)			(30.397)		

Table 6-7: Picture Developability—Level Sum and Average

Level / Factor	Level Sum			Average		
	1	2	3	1	2	3
A	– 286.83	– 292.18	—	– 31.87	– 32.46	—
B	– 175.09	– 197.12	–206.80	–29.18	– 32.85	– 34.47
C	– 196.03	– 190.34	– 192.64	– 32.67	– 31.72	– 32.11
D	– 194.28	– 193.29	– 191.44	– 32.38	– 32.21	– 31.91
E	– 198.42	– 193.01	– 187.58	– 33.07	– 32.17	– 31.26
F	– 196.80	– 192.52	– 189.69	– 32.80	– 32.09	– 31.62
G	– 200.97	– 191.65	– 186.38	– 33.50	– 31.94	– 31.06
H	– 196.26	– 191.21	– 191.53	– 32.71	– 31.87	– 31.92
(*T*)	(– 579.004)			(– 32.167)		

Step 15. Drawing the factorial effects chart

The averages summarized in *Table 6-6* and *Table 6-7* were used to draw the factorial effects chart. The results are shown in *Figure 6-5* and *Figure 6-6*.

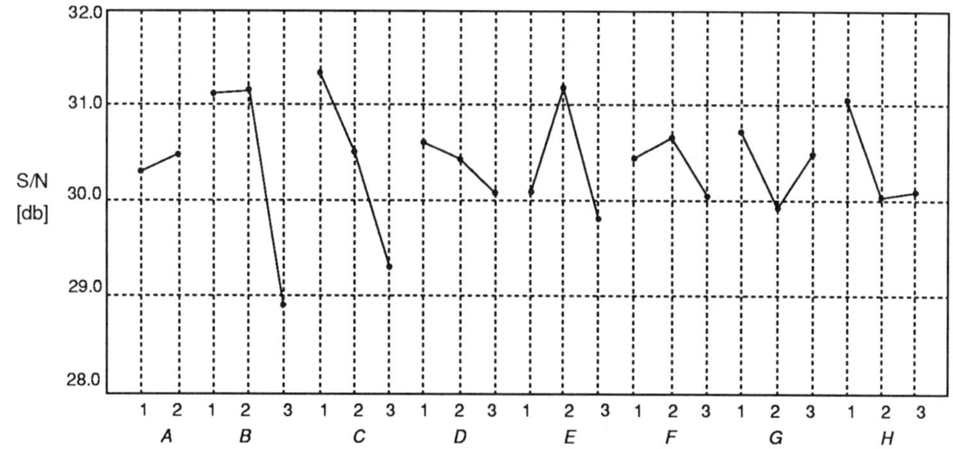

Figure 6-5: Factor Effects — Adhesiveness

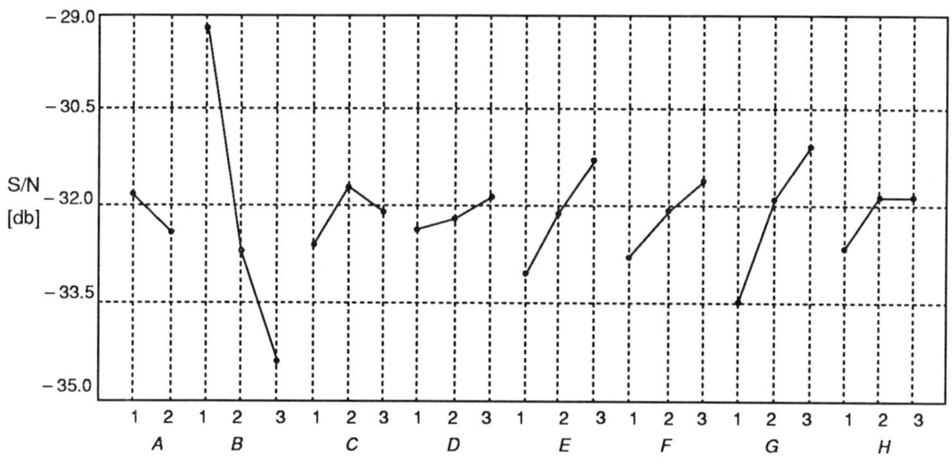

Figure 6-6: Factor Effects — Picture Developability

Step 16. Optimum condition estimation

The effective factors for adhesiveness are *B*, *C*, and *E*, and their optimum levels are B_2, C_1, and E_2. The effective factors for development are *B*, *E*, and *G*, and their optimum levels are B_1, E_3, and G_3. The effect of levels B_1 and B_2 of factor B for adhesiveness is about the same (31.15 and 31.16, respectively). As for factor *E*, E_3 is selected because picture developability is the more important characteristic. Therefore, the final optimum condition is B_1, C_1, E_3, and G_3. The levels of factors that are not effective, i.e., *A*, *D*, *F*, and *H*, were chosen as A_1, D_2, F_2, and H_2 by considering cost, productivity, and quality.

The estimate of the optimum condition ($\hat{\mu}$) was calculated as follows:
Estimate from the factors effective for adhesiveness

$$\hat{\mu}(B_1C_1E_3) = \overline{B}_1 + \overline{C}_1 + \overline{E}_3 - 2 \times \overline{T}$$
$$= 31.15 + 31.38 + 29.84 - 2 \times 30.397$$
$$= 31.576 \quad [\text{db}]$$

Estimate from the factors effective for picture developability

$$\hat{\mu}(B_1E_3G_3) = \overline{B}_1 + \overline{E}_3 + \overline{G}_3 - 2 \times \overline{T}$$
$$= -29.18 - 31.26 - 31.06 - 2 \times (-32.167)$$
$$= -27.166 \quad [\text{db}]$$

The S/N ratio is reverse-transformed to obtain the adhesiveness (m_1) and the quantity of remaining resin after development (m_2):

$$31.576 = -10\log\frac{1}{m_1^2} \qquad m_1 = 37.914$$

$$-27.166 = -10\log\frac{1}{m_2^2} \qquad m_2 = 22.819$$

Because these results clear the initial target, i.e., above 35 and below 25 respectively, the confirmatory trial was conducted.

The results were as follows:

	Standard Condition	**Accelerated Condition**	**S/N Ratio (db)**
Adhesiveness	38	37	31.478
Development	21	24	-27.063

From these results, the condition for the production line was determined to be A_1, B_1, C_1, D_2, E_3, F_2, G_3, and H_2.

6-2-2. Discussion of the Study

Q. Why is the level of factor *A* different depending on the level of factor *B*?

M. If the level of factor *A* is determined independently of the level of factor *B*, the interaction is expected to affect the result. Therefore, the levels of factor *A* are set at two, i.e., large and small, and different numbers are assigned to the levels of *A* for different levels of factor *B*. This is similar to the case of additives or catalysts, where the quantity is determined as the ratio or proportion of a certain raw material. Please see Section 5-4-7 for more on this subject.

Q. Are there any other cases where contradicting characteristics must be obtained under the same condition as in this example?

M. There are numerous examples. They include:

- A tape with good adhesiveness that can be easily peeled off

- A lid of a can that closes tightly but can be easily opened
- A pencil with a solid black color that can be easily erased.

There are many other examples.

Q. What is the key point about the analysis of these examples?

M. First classify the contradicting characteristics into STB and LTB data and determine the optimum condition for each characteristic. Then adjust these optimum conditions on the basis of the importance of the characteristics in order to derive the final optimum condition.

Q. How do we determine the enforcement condition as shown in *Table 6-5*?

M. The **accelerated condition** must satisfy the following:

1. It must be possible to make an evaluation in a short period of time.
2. The deterioration and destruction modes must be the same as those on the factory floor and in the marketplace.

Q. I can understand the importance of the first requirement, but how do we determine the level of acceleration?

M. Although this depends on engineering technology, the level of acceleration must be such that one-third to two-thirds of the samples will have a problem provided the failure mode is the same.

Q. Sometimes, the accelerated test is conducted until the sample becomes nonfunctional.

M. The worst way to determine the optimum condition is the practice of making judgments based on the point at which the sample becomes nonfunctional. It is true that such a test is done to judge whether or not the product should be shipped out. But, because the purpose of the test is to determine the optimum condition, this judgment should be based on continuous data.

Q. How did you select the accelerated condition in this example?

M. The accelerated condition of adhesiveness is set at 20% reduction in exposure quantity. This could have been 50% because there is not much difference between the standard and enforcement conditions. Although the accelerated condition of picture developability was set at four-fifths solution, this could have been three-fifths to one-half. These conditions correspond to the worse conditions in the marketplace.

The accelerated condition must be examined differently for different purposes because it is necessary to have technological know-how in order to evaluate the condition.

6-3. Improvement of Chemical Plating Quantity Control and Variability of an Electronic Part (Parameter Design)

This research relates to the chemical plating equipment and the process design of an electronic part. Chemical plating is done on a portion of the part, and 50 part units are simultaneously processed in a basket. The target value of the quantity of chemical plating is 200, and it is necessary to design the equipment and the process that will give all part units in the basket a uniform plated film thickness.

The parameter design method is effective if we want to adjust the output to the target value and reduce variability at the same time. The Steps of this method are as follows:

1. Select the combination of levels that will minimize variability (based on the S/N ratio).
2. Adjust the output given by the level combination of (1) to the target value using factors that can change the mean without affecting the S/N ratio.

A key feature of this method is that it allows us to obtain reduced variability and target output using only combinations of the assigned factors and their levels on the basis of a small-scale batch test.

6-3-1. Research Content

Step 1. Factor listing

Personnel involved (equipment operators and engineers) were assembled to list the factors.

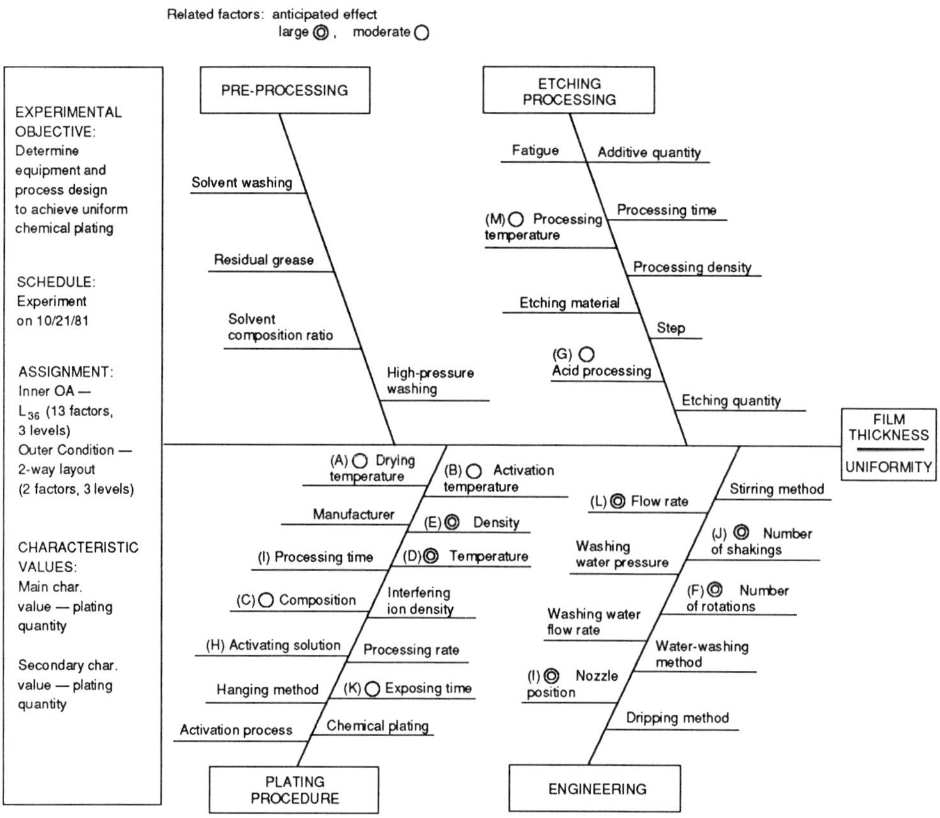

Figure 6-7: Uniformity of Chemical Plating

Steps 2–5. Factor classification — determination of level

The listed factors were classified into control, indicative, block, and noise factors. The effects of the control factors were evaluated. They were marked with a double circle if they were considered to have a significant relation to the output and with a single circle for a moderate relation. The factors are summarized in *Figure 6-7*.

Steps 6–9. Selection of orthogonal arrays and preparation of experiment instructions

There are two possible orthogonal arrays in the parameter design. An L_{36} was selected because the number of factors needed to be tested was more than 10. Thirteen specific columns of the L_{36} were used for the parameter design. It is possible to obtain main effects free of interactions because the interactions influence each column of the L_{36} equally. As for the outer noise condition, two samples from the edge and middle of the basket were selected and their front and back surfaces were measured.

Usually, the outer condition is determined by compounding the noise factors, but in this case, the temperature, density, and diffusion are different for the middle and the edge of the basket. Therefore, two samples from the edge and the middle of the basket were picked out and four points were measured, including the plating quantity of the front and back surfaces. Because the sample is an asymmetrical plate, its front and back surfaces were used as the noise factor. In other words, it is necessary to obtain parts that are uniformly plated on the front and back surfaces, regardless of their position in the basket.

Table 6-8 (below) shows the factors and levels. *Table 6-9* (next page) shows the assignment. The experiment instructions are omitted.

Table 6-8: Factor and Level

Symbol	Factor Name	Level		
		1	2	3
A	Drying temperature (°C)	40	60	80
B	Activation temperature (°C)	20	30	40
C	Composition	1.2/0.8	1/1	0.8/1.2
D	Temperature (°C)	80	70	60
E	Density (%)	20	30	40
F	Number of rotations (rpm)	3	6	9
G	Acid processing (sec)	10	15	20
H	Activating solution (g/l)	10	20	30
I	Activating time (min.)	1	3	7
J	Number of shakings	10	20	30
K	Exposing time	1	2	3
L	Flow rate	0.5	1.0	1.5
M	Processing temperature (°C)	40	60	80

Table 6-9a: Assignment and Data

Factor	*A*	*B*	*C*	*D*	*E*	*F*	*G*	*H*	*I*	*J*	*K*	*L*	*M*	Observed (Raw) Data Edge of the Basket	
Col. No.	1	2	3	4	5	6	7	8	9	10	11	12	13	Front	Back
1	1	1	1	1	1	1	1	1	1	1	1	1	1	108	110
2	2	2	2	2	2	2	2	2	2	2	2	2	1	75	72
3	3	3	3	3	3	3	3	3	3	3	3	3	1	143	139
4	1	1	1	1	2	2	2	2	3	3	3	3	1	263	253
5	2	2	2	2	3	3	3	3	1	1	1	1	1	89	84
6	3	3	3	3	1	1	1	1	2	2	2	2	1	73	72
7	1	1	2	3	1	2	3	3	1	2	2	3	1	68	62
8	2	2	3	1	2	3	1	1	2	3	3	1	1	418	398
9	3	3	1	2	3	1	2	2	3	1	1	2	1	92	92
10	1	1	3	2	1	3	2	3	2	1	3	2	1	104	103
11	2	2	1	3	2	1	3	1	3	2	1	3	1	76	72
12	3	3	2	1	3	2	1	2	1	3	2	1	1	224	204
13	1	2	3	1	3	2	1	3	3	2	1	2	2	600	599
14	2	3	1	2	1	3	2	1	1	3	2	3	2	58	60
15	3	1	2	3	2	1	3	2	2	1	3	1	2	81	74
16	1	2	3	2	1	1	3	2	3	3	2	1	2	70	64
17	2	3	1	3	2	2	1	3	1	1	3	2	2	86	90
18	3	1	2	1	3	3	2	1	2	2	1	3	2	324	327
19	1	2	1	3	3	3	1	2	2	1	2	3	2	114	105
20	2	3	2	1	1	1	2	3	3	2	3	1	2	122	136
21	3	1	3	2	2	2	3	1	1	3	1	2	2	90	85
22	1	2	2	3	3	1	2	1	1	3	3	2	2	57	55
23	2	3	3	1	1	2	3	2	2	1	1	3	2	214	213
24	3	1	1	2	2	3	1	3	3	2	2	1	2	73	82
25	1	3	2	1	2	3	3	1	3	1	2	2	3	428	418
26	2	1	3	2	3	1	1	2	1	2	3	3	3	115	116
27	3	2	1	3	1	2	2	3	2	3	1	1	3	54	51
28	1	3	2	2	2	1	1	3	2	3	1	3	3	62	64
29	2	1	3	3	3	2	2	1	3	1	2	1	3	151	189
30	3	2	1	1	1	3	3	2	1	2	3	2	3	80	78
31	1	3	3	3	2	3	2	2	1	2	1	1	3	83	83
32	2	1	1	1	3	1	3	3	2	3	2	2	3	153	155
33	3	2	2	2	1	2	1	1	3	1	3	3	3	73	69
34	1	3	1	2	3	2	3	1	2	2	3	1	3	77	76
35	2	1	2	3	1	3	1	2	3	3	1	2	3	53	59
36	3	2	3	1	2	1	2	3	1	1	2	3	3	291	374

Table 6-9b: Assignment and Data

Observed Data (Raw Data)		**Transformed Data**	
Middle of Basket		Sensitivity *S* [db]	S/N Ratio η [db]
Front	Back		
109	89	46.361	20.302
74	76	43.434	32.765
137	140	48.928	34.941
258	261	54.278	35.498
87	94	44.959	26.465
72	76	43.317	31.748
66	70	42.477	25.785
415	397	58.212	31.330
102	93	45.552	25.804
94	95	45.933	25.542
76	81	43.665	26.313
191	192	52.160	22.412
580	578	61.427	33.915
51	53	40.906	22.408
78	80	43.890	28.053
70	67	42.639	27.449
78	83	44.532	24.427
317	311	56.117	32.971
103	103	46.457	26.119
119	121	47.924	24.093
85	88	44.811	31.008
58	61	41.252	27.269
208	235	52.770	25.192
73	74	43.579	24.766
487	459	59.046	23.105
108	110	47.024	29.265
54	54	40.547	31.004
66	64	42.144	31.864
176	171	50.718	20.728
73	68	43.493	22.855
68	80	43.918	20.812
144	131	49.293	22.490
70	75	43.137	28.314
74	73	43.522	32.276
50	57	40.788	22.652
280	276	54.971	31.346

Steps 10–11. Data transformation

There were no missing data in the samples. The raw data were transformed into S_m and η.

S_m is called the general average variation or the sensitivity, and it expresses the impact of the factor level changes on the output. In this example, it is the change in plating quantity when the level is changed. The **general average, S_m,** is obtained by the following formula:

$$S_m = 10\log\frac{1}{n}(\Sigma y_i)^2 \quad [\text{db}]$$

S_m for the first experiment was calculated as follows:

$$S_m = 10\log\frac{1}{4}(108+110+109+89)^2$$
$$= 46.361 \quad [\text{db}]$$

η is called the S/N ratio and expressed as follows:

$$\text{S / N ratio} \propto \frac{\text{Signal}}{\text{Noise}} = \frac{m^2}{\sigma^2} = \frac{\text{Plating Quantity}}{\text{Variation}}$$

The S/N ratio enables evaluation of the effect of noise against the signal. In this example, the degree of variability when the plating quantity is changed is evaluated. The S/N ratio, η, is calculated according to the following formulae:

$$\eta = 10\log\frac{\frac{1}{n}(S_m - V_e)}{V_e} \quad [\text{db}]$$

where n: number of data, and

$$T = \sum_{i=1}^{n} y_i^2$$

$$SM = \frac{1}{n}T^2$$

$$X = \sum\left(y_i^2\right)$$

$$S_e = X - SM$$

$$V_e = \frac{1}{n-1}\times S_e$$

η_1 for the first experiment is calculated as follows:

$$T = 108 + 110 + 109 + 89 = 416$$

$$SM = \frac{1}{4}(416)^2 = 43264$$

$$X = (108)^2 + (110)^2 + (109)^2 + (89)^2 = 43566$$

$$S_e = (43566 - 43264) = 302$$

$$V_e = \frac{302}{4-1} = 100.67$$

$$\eta_1 = 10\log\frac{\frac{1}{4}(43264-100.67)}{100.67} = 20.302 \quad [\text{db}]$$

The S/N ratios for the rest of the experiments were similarly calculated. The results are summarized in *Table 6-9b*.

Steps 12–13. Level sum and average

The level sum and the average were calculated next.

For example, the level sum and the average of sensitivity (S) for factor A were calculated as follows:

Level sum of A_1 $\quad A_1 = 46.361 + 54.278 + \cdots + 43.522 = 569.54$

Level sum of A_2 $\quad A_2 = 43.434 + 44.959 + \cdots + 40.788 = 564.23$

Level sum of A_3 $\quad A_3 = 48.928 + 43.317 + \cdots + 54.971 = 560.50$

The average is obtained by dividing the level sum by the number of data:

$$\overline{A}_1 = \frac{A_1}{12} = \frac{569.54}{12} = 47.46 \text{ db}$$

$$\overline{A}_2 = \frac{A_2}{12} = \frac{564.23}{12} = 47.02 \text{ db}$$

$$\overline{A}_3 = \frac{A_3}{12} = \frac{560.50}{12} = 46.71 \text{ db}$$

The level sum and the average for other factors and S/N ratios were calculated similarly. *Table 6-10* and *Table 6-11* show the results.

Table 6-10: Sensitivity (S db)
Level Sum and Average

Level / Factor	Level Sum			Average		
	1	2	3	1	2	3
A	569.54	564.23	560.50	47.46	47.02	46.71
B	565.27	564.28	564.72	47.11	47.02	47.06
C	542.28	557.33	594.67	45.19	46.44	49.56
D	636.05	527.64	530.58	53.00	43.97	44.21
E	530.29	576.48	587.50	44.19	48.04	48.96
F	548.03	573.81	572.43	45.67	47.82	47.70
G	569.23	565.55	559.49	47.44	47.13	46.62
H	571.06	556.49	566.71	47.59	46.37	47.23
I	546.86	565.73	581.68	45.57	47.14	48.47
J	578.42	559.90	555.96	48.20	46.66	46.33
K	563.06	569.09	562.13	46.92	47.42	46.84
L	558.43	562.88	572.96	46.54	46.91	47.75
M	569.28	566.39	558.60	47.44	47.20	46.55
(T)	(1697.27)			(47.063)		

Table 6-11: S/N Ratio (η)
Level Sum and Average

Level / Factor	Level Sum 1	Level Sum 2	Level Sum 3	Average 1	Average 2	Average 3
A	329.94	308.13	345.22	27.49	25.68	28.77
B	319.06	345.14	319.08	26.59	28.76	26.59
C	314.26	325.75	343.28	26.19	27.15	28.61
D	325.51	337.93	319.85	27.13	28.16	26.65
E	307.34	341.29	334.66	25.61	28.44	27.89
F	326.00	343.32	313.97	27.17	28.61	26.16
G	327.11	330.24	325.93	27.26	27.52	27.16
H	327.77	318.88	336.64	27.31	26.57	28.05
I	304.35	351.35	327.58	25.36	29.28	27.30
J	305.40	337.56	340.33	25.45	28.13	28.36
K	328.30	311.12	343.86	27.36	25.93	28.66
L	309.69	323.58	350.02	25.81	26.97	29.17
M	338.91	327.67	316.71	28.24	27.31	26.39
(*T*)	(983.286)			(27.314)		

Step 15. Drawing the factorial effects chart

Based on the averages summarized in *Table 6-10* and *Table 6-11*, the factorial effects charts for the general average (*S*) and the S/N ratio (η), were drawn, as shown in *Figure 6-8* and *Figure 6-9*.

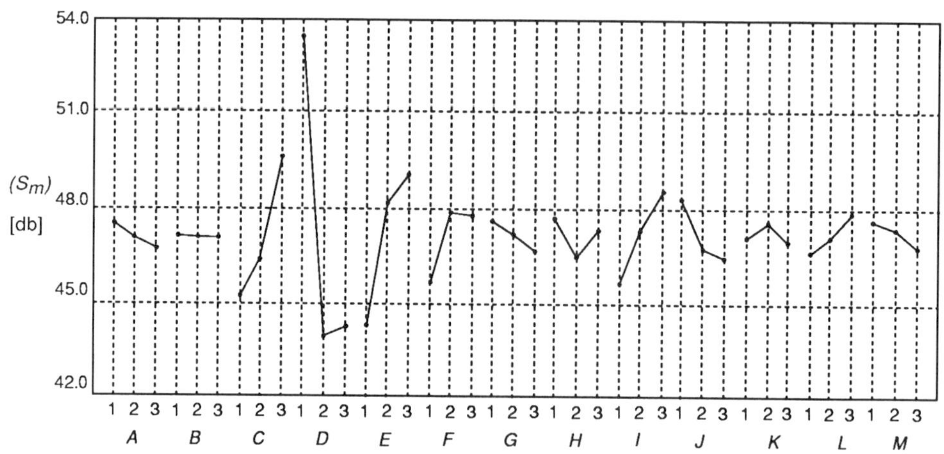

Figure 6-8: Factorial Effects for Sensitivity (*S*)

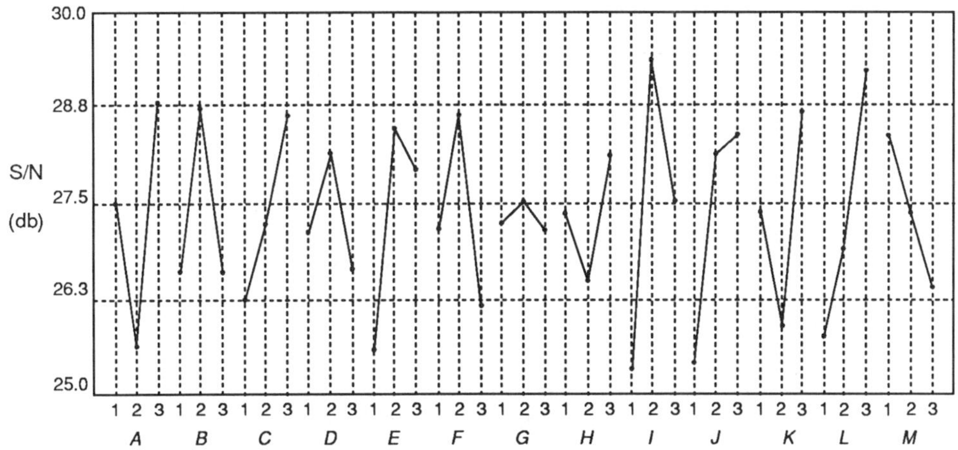

Figure 6-9: Factorial Effects for S/N Ratio (η)

Step 16. Optimum condition estimate

The optimum condition in this research was obtained according to the following method:

1. Select the combination of levels that minimizes the variability (the combination with the maximum η).
2. Adjust the difference between the target output and the output of the level combination obtained in (1) with factors that do not influence η, but that have a significant effect on *S* (adjustment using *S*).

Effective factors and their levels for the S/N ratio (η) and the sensitivity (*S*) are summarized in *Table 6-12*. According to *Figure 6-9*, the factorial effects chart for η, factors *A* and *I* have a significant effect, and factors *B, C, E, F, J, K,* and *L* have a moderate effect. Their optimum levels are A_3, B_2, C_3, E_2, F_2, I_2, J_3, K_3, and L_3. The estimate ($\hat{\mu}$) of the optimum condition is calculated as follows:

$$\hat{\mu}_{(\eta)} = \overline{A}_3 + \overline{B}_2 + \overline{C}_3 + \overline{E}_2 + \overline{F}_2 + \overline{I}_2 + \overline{J}_3 + \overline{K}_3 + \overline{L}_3 - 8 \times \overline{T}$$

$$= 28.77 + 28.76 + 28.61 + 28.44 + 28.61 + 29.28 + 28.36 + 28.66$$

$$+ 29.17 - 8 \times 27.314$$

$$= 40.148 \quad [\text{db}]$$

From *Figure 6-8*, the effective factors for the general average are *C, D,* and *E*. *D* is not related to η and is effective only on *S*. *C* and E are fixed at C_3 and E_2, respectively, from the S/N ratio (η). Therefore, the S/N ratio (η), for the quantity of chemical plating is estimated for the combinations of D_1, D_2, and D_3 and C_3 and E_2.

However, it was expected that the estimate of the S/N ratio (η) based on 9 out of 13 factors used in the optimum condition will be too high. Therefore, the estimate was based on the four factors whose S/N ratio was the highest.

$$\hat{\mu}_{(\eta)} = \overline{A}_3 + \overline{B}_2 + \overline{I}_2 + \overline{L}_3 - 3 \times \overline{T}$$

$$= 28.77 + 28.76 + 29.28 + 29.17 - 3 \times 27.314$$

$$= 34.038 \quad [\text{db}]$$

The estimate of sensitivity (S) for $\hat{\mu}_\eta$ is based on factors C, D, and E, which are effective for sensitivity. C and E are fixed at C_3 and E_2 since they are effective for the S/N ratio. $\hat{\mu}_{(m1)}(C_3D_1E_2)$, $\hat{\mu}_{(m2)}(C_3D_2E_2)$, and $\hat{\mu}_{(m3)}(C_3D_3E_2)$ are calculated as follows:

$$\hat{\mu}_{(m_1)}(C_3D_1E_2) = 49.56 + 53.00 + 48.04 - 2\times 47.06 = 56.48 \quad [\text{db}]$$

$$\hat{\mu}_{(m_2)}(C_3D_2E_2) = 49.56 + 43.97 + 48.04 - 2\times 47.06 = 47.45 \quad [\text{db}]$$

$$\hat{\mu}_{(m_3)}(C_3D_3E_2) = 49.56 + 44.21 + 48.04 - 2\times 47.06 = 47.69 \quad [\text{db}]$$

The quantity of chemical plating (m_1, m_2, m_3) is estimated as follows:

$m_1 = 333.4$ ($56.48 = \log(1/4)(4m_1)^2$)
$m_2 = 117.9$ ($47.45 = \log(1/4)(4m_2)^2$)
$m_3 = 121.2$ ($47.69 = \log(1/4)(4m_3)^2$)

Next, because the target value of the quantity of chemical plating is 200, the precision test was carried out by putting 2°C intervals between D_2 and D_1 and keeping A_3, B_2, C_3, E_2, F_2, I_2, J_3, K_3, and L_3. The results are shown in *Figure 6-10*. The interval between D_2 and D_3 was not tested because it does not give the target value, 200.

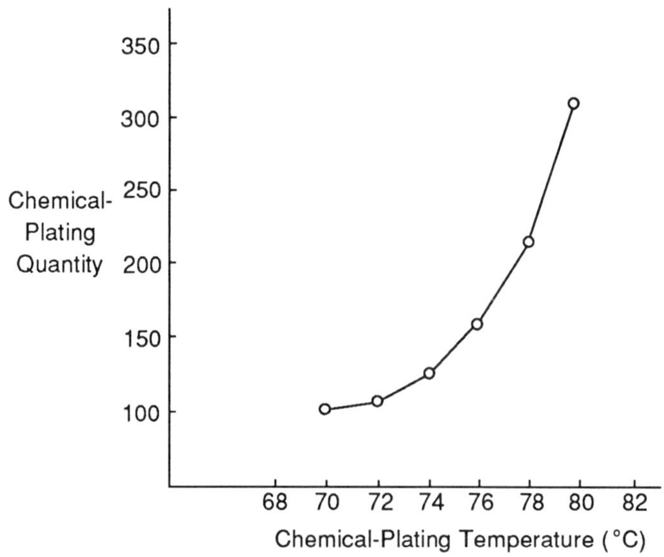

Figure 6-10: Chemical-Plating Quantity and Temperature

Based on *Figure 6-10*, in order to achieve a chemical-plating quantity of 200 the temperature was set at 77.5°C. The confirmatory trial with 77.5°C gave samples with chemical-plating quantities of 206, 199, 198, and 201.

Table 6-12: Effective Factors, Levels, and Optimum Condition

Factor	Level	S/N Ratio (η)	Sensitivity (*S*)	Optimum Level (Total Judgment)
A	1			
	2			
	3	◎		○
B	1			
	2	○		○
	3			
C	1			
	2			
	3	○	↓	○ (77.5 °C)
D	1		⇧	
	2			
	3			
E	1			
	2	○		
	3	(○)	↓	○
F	1			
	2	○		○
	3			
G	1			
	2			
	3			
H	1			
	2			
	3			
I	1			
	2	◎		○
	3			
J	1			
	2	(○)		
	3	○		○
K	1			
	2			
	3	○		○
L	1			
	2			
	3	○		○
M	1			
	2			
	3			

◎ Large effect

○ Some effect

⇨ Large effect

→ Some effect

The direction of the arrow indicates the direction of plating-quantity increase

Their average was 201.

$$T = 206 + 199 + 198 + 201 = 804$$

$$SM = \frac{1}{4}(804)^2 = 161604$$

$$X = (206)^2 + (199)^2 + (198)^2 + (201)^2 = 161642$$

$$S_e = X - SM = 161642 - 161604 = 38$$

$$V_e = \frac{38}{3} = 12.667$$

From the preceding data, the estimate of sensitivity and the S/N ratio are obtained in the following way:

$$\text{Sensitivity} \quad S = 10\log\frac{1}{n}(T)^2 = 10\log\frac{1}{4}(804)^2 = 52.084 \quad [\text{db}]$$

$$\text{S/N ratio } \eta = 10\log\frac{\frac{1}{4}(SM - V_e)}{V_e} = 10\log\frac{\frac{1}{4}(161604 - 12.667)}{12.667}$$

$$= 35.037 \quad [\text{db}]$$

This condition enabled achievement of the target value, 200, with small variability.

6-3-2. Discussion of the Study

Q. I had not heard of a design method in which we first obtain the smallest variability and then adjust the output level.

M. In the traditional method, factor levels that have a significant effect on the general average are used to adjust the output to the target. Variability is automatically determined in this process. This is not design, rather it is a step in design called the adjustment method. Design is finding an inexpensive production process whose output coincides with the target value, with small variability, and in a short period of time.

Q. Usually, reduction in variability is achieved by obtaining a uniform temperature or density that has locality (distribution). Because variability is the result of a local phenomenon, the countermeasure should be to achieve a uniform condition.

M. That is wrong thinking. Tremendous investments are required for eliminating the locality of mass production equipment.

For example, when defects were caused by the distribution of internal furnace temperatures at INAX Co., Ltd., the defects were eliminated by changing the quantity of additives instead of achieving a uniform temperature distribution within the furnace by controlling the temperature. This example is important because the defects were eliminated by adjusting the levels of factors not related to the temperature distribution, which was the actual cause of defects. It is possible to reduce the impact of variability on the output by changing the level of other inexpensive factors without actually controlling the cause of variability.

Q. Do you have another example?

M. Yes. The thickness of plated metal film differs according to the distribution of electrical current. If current distribution were to be made uniform, it would be necessary to have many supplementary electrodes. But the same objective can be achieved by using a high density, high temperature, and high stirring flow rate. Direct control of the cause is not a good solution because of the high cost involved.

Q. Was the outer condition appropriately taken in this example? I think it is better to take slightly different levels of the factors assigned to the inner orthogonal array.

M. That is quite right. However, if we do that, the number of experiments becomes 32×4, which is impractical. Because such conditions as density, temperature, and stirring flow rate are different at the center and the edge of the basket, as well as for the front and back surfaces of the sample, the outer condition was determined by taking samples from the middle and the side of the basket and measuring both surfaces.

Q. There was no significant factor for the S/N ratio while nine factors have about the same moderate effect.

M. Unfortunately, there was no significant factor. In such a case, the estimate of the optimum condition must be done with about four factors, one-third of the moderate effect factors, because the estimate will be overvalued if all factors are used.

Q. Factor D is quite suitable as an adjustment factor because it has a significant impact on the general average (S) but it does not influence the S/N ratio (η).

M. The existence of factor D was the key to success in this research. We are not usually this lucky. In order to find this kind of convenient factor, we need to assign main effect factors to all columns and to take wide level intervals. An L_{36}, which has more assignment columns, is better than an L_{18} for finding such factors.

Q. What should we do if the level combination we obtain to minimize the variability still does not sufficiently reduce the variability?

M. In such a case, it is necessary to control the variation of the selected level setting. This leads to a higher cost because it attempts to control the cause of variability. With tolerance design we can determine which factors need to be controlled and to what degree. This will be discussed in the chapter on tolerance design, which follows later.

7

SPECIALIZED FACTOR ASSIGNMENT TECHNIQUES

When we assign multiple factors and levels in an experiment to an orthogonal array, for efficient experimentation we will need various techniques in addition to the standard assignment technique. Sometimes, we may want to assign four or three levels in a two-level orthogonal array, or two levels in a three level orthogonal array. These correspond to the **multi-level arrangement technique**.

There is another technique called the **combination technique**. This technique combines two factors with either strong, weak, or medium effects into one factor when these two factors have similar effects. The original two factors are judged to be noneffective if the different levels of the combined factor do not show a different result.

The **nested-factor technique** is used when heterogeneous levels are taken for one factor. For example, when the alumite and chemical chrome-plating methods are used to cure aluminum, the alumite condition and the chrome-plating condition should not appear on the same sample. The nested-factor method is used to solve this kind of problem.

This chapter explains the multi-level arrangement, the combination technique, and the nested-factor method and provides examples of their application.

7-1. Four- and Eight-Level Arrangement in the Two-Level System

In order to arrange **four levels** in two-level system orthogonal arrays (this can be done with an L_8, L_{16}, L_{32}, and others but not an L_{12}), two arbitrary columns and their interaction must be sacrificed. The interaction column can be found by reading the number in the triangular table, which has the interaction between the main effect columns, labelled "Interaction Between Two Columns." This means that a total of three columns are sacrificed, and two arbitrary columns are picked up. After elimination of the third column, the numbers 1, 2, 3, and 4 are assigned to the combinations (1, 1), (1, 2), (2, 1), (2, 2) of the levels of these two columns.

Consider an experiment where columns 1, 2, and 3 are used for factor *A* with four levels, and the rest of the columns are assigned to four other two-level factors: *B*, *C*, *D*, and *E*. The concept of this arrangement is shown in Figure 7-1, where a closed circle indicates the multi-level, and levels one and two of columns 1 and 2 are used to create four levels. Columns 1 and 2 are expressed by points (circles) and the interaction between them is expressed by the line between the points. From the intersection of the first and second columns in the triangular table, it is clear that the interaction is given by column 3. Other factors, *B*, *C*, *D*, and *E*, are expressed with points because main effects alone are derived for these factors. The figure that shows these relationships using points and edges is called the **linear graph**. The linear graph has factor names (*A*, *B*, ...), lines, points, and numbers indicating the corresponding columns in the orthogonal array. *Table 7-1* shows the orthogonal array before and after the four-level assignment. Typical linear graphs are shown in the appendix, including those for interactions between two columns.

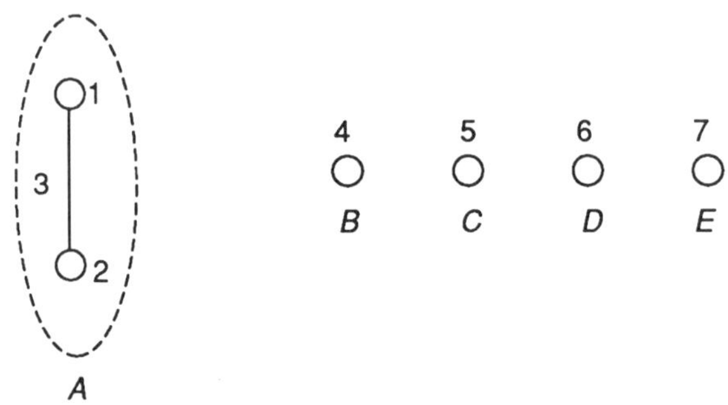

Figure 7-1: Linear Graph L_8 ($4^1 \times 2^4$)

Table 7-1: The Orthogonal Array Before and After Four -Level Arrangement

	L_8 (2^7)								**L_8 ($4^1 \times 2^4$)**				
Column / **No.**	1	2	3	4	5	6	7	**Factor** / **Column** / **No.**	*A* (1 2 3)	*B* (4)	*C* (5)	*D* (6)	*E* (7)
1	1	1	1	1	1	1	1	1	1	1	1	1	1
2	1	1	1	2	2	2	2	2	1	2	2	2	2
3	1	2	2	1	1	2	2	3	2	1	1	2	2
4	1	2	2	2	2	1	1	4	2	2	2	1	1
5	2	1	2	1	2	1	2	5	3	1	2	1	2
6	2	1	2	2	1	2	1	6	3	2	1	2	1
7	2	2	1	1	2	2	1	7	4	1	2	2	1
8	2	2	1	2	1	1	2	8	4	2	1	1	2

Standard Array | Array After Multi-Level Factor

In order to arrange **eight levels** in the two-level system, a triangle (which has one line for the interaction column connecting a vertex and an opposite baseline) can be used. In the case of an L_{16}, columns 1, 2, and 4 are assigned to the vertices of the triangle, and the three columns that will be sacrificed are: column 1, which corresponds to one top; column 6, which corresponds to the baseline; and column 7, which corresponds to the interaction between columns 1 and 6. The eight levels of one factor are given by the eight combinations of levels 1 and 2 of columns 1, 2, and 4. This is shown in *Table 7-2*. The linear graph and the final orthogonal array are shown in *Figure 7-2* and *Table 7-3*, respectively. The factors assigned in the final orthogonal array are factor *A*, with eight levels, and factors *B* through *I*, with two levels each.

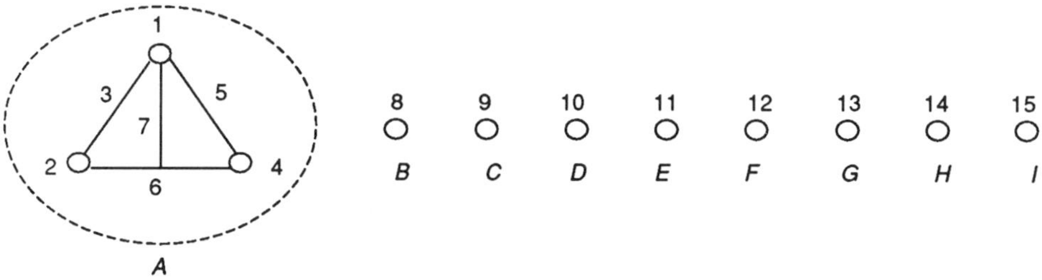

Figure 7-2: Linear Graph of Eight Level Factor *A* and Two Level Factors *B* ~ *I* (L_{16} $(8^1 \times 2^8)$)

Table 7-2: Eight-Level Arrangement in the Two-Level System

Column 1	Column 2	Column 4	*A*
1	1	1	→1
1	1	2	→2
1	2	1	→3
1	2	2	→4
2	1	1	→5
2	1	2	→6
2	2	1	→7
2	2	2	→8

Table 7-3: Orthogonal Array $L_{16}(8^1 x 2^8)$ Created From Orthogonal Array $L_{16}(2^{15})$

Original Column	1 ~7	8	9	10	11	12	13	14	15
No.	*A*	*B*	*C*	*D*	*E*	*F*	*G*	*H*	*I*
1	1	1	1	1	1	1	1	1	1
2	1	2	2	2	2	2	2	2	2
3	2	1	1	1	1	2	2	2	2
4	2	2	2	2	2	1	1	1	1
5	3	1	1	2	2	1	1	2	2
6	3	2	2	1	1	2	2	1	1
7	4	1	1	2	2	2	2	1	1
8	4	2	2	1	1	1	1	2	2
9	5	1	2	1	2	1	2	1	2
10	5	2	1	2	1	2	1	2	1
11	6	1	2	1	2	2	1	2	1
12	6	2	1	2	1	1	2	1	2
13	7	1	2	2	1	1	2	2	1
14	7	2	1	1	2	2	1	1	2
15	8	1	2	2	1	2	1	1	2
16	8	2	1	1	2	1	2	2	1

7-2 Three-Level Arrangement in the Two-Level System (Dummy-Level Technique)

The previous section explained the four-level arrangement in the two level system. This section explains the arrangement of a three-level factor in two-level system orthogonal arrays using the **dummy-level technique**. First, the four levels are arranged according to the previous method. In order to change two levels of factor *A* into three levels in the right hand table in *Table 7-1*, the most important (standard) level must be duplicated. The method of duplicating a level is called the dummy-level technique. If the second level, A_2, is to be duplicated as the fourth level, the levels become:

$$A_1 = A_1, A_2 = A_2, A_3 = A_3, A_4 = A'_2$$

The duplicated level is indicated by a prime symbol ('). The corresponding linear graph and orthogonal array are shown in *Figure 7-3* and *Table 7-4*. The level sum of A_2 and A'_2 is calculated by addition, and the number of replications are obtained from the number of data under A_2 and A'_2. The first order error variance (Se_1) for the analysis of variance is derived from A_2 and A'_2. The mean of A_2 is given by the average of $\bar{A}_2$ and $\bar{A}'_2$.

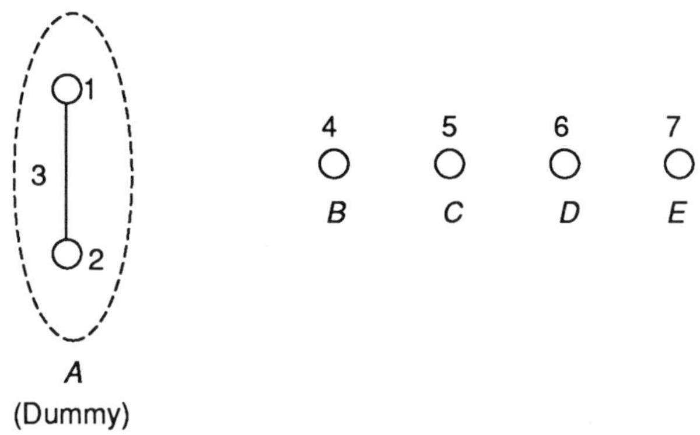

Figure 7-3: Linear Graph for Dummy Treatment

Table 7-4: Three-Level Assignment Based on the Dummy-Level Technique

Factor / Column / No.	A (1 2 3)	B (4)	C (5)	D (6)	E (7)
1	1	1	1	1	1
2	1	2	2	2	2
3	2	1	1	2	2
4	2	2	2	1	1
5	3	1	2	1	2
6	3	2	1	2	1
7	2'	1	2	2	1
8	2'	2	1	1	2

7-3 Three-Level Assignment in the Two-Level System: (Idle-Column Technique)

The dummy-level technique allows assignment of only one factor with three levels in an L_8. This is because when we create the third level of the second factor, it interacts with the already created column 3. As shown earlier, three levels are preferred in experimentation. This section explains the idle-column technique of multi-level arrangement, which reduces the number of sacrificed columns, as compared to the dummy-level technique. Consider the case where factors *A* and *B*, with three levels each, and factors *C* and *D*, with two levels each, are assigned to an L_8.

Columns 1, 2, and 3 correspond to factor *A*. Columns 1, 4, and 5 correspond to factor *B*. Columns 6 and 7 correspond to factors *C* and *D*, respectively. Because column 1 is used for both factors *A* and *B*, we cannot apply the method of creating three levels from the four-level system based on the dummy-level technique. In such a case, three levels can be created by reading the levels of factors *A* and *B*, corresponding to level 1 of common column 1, as 1 and 2, and by reading the levels of factors A and B, corresponding to the level 2 of common column 1, as 2 and 3 instead of 1 and 2. This common column is called ***the* idle column,** and this three-level arrangement is called, the **idle-column technique**.

The linear graph and the orthogonal array for this example are shown in *Figure 7-4* and *Table 7-5*, respectively. The third level of factors A and B is created from columns 2 and 4 for the corresponding levels 1 and 2 of the junction column. The factors with three levels used in the idle-column technique are not orthogonal for levels 1 and 2 of the idle column. For example, there is no combination of A_1 and B_3. Therefore, level sums, averages, factorial effects charts, and variation calculations are done separately for levels 1 and 2 of the idle column. However, the factorial effects are adjusted so that level 2 (**the reference level**) of each factor, which is common to levels 1 and 2 of the idle column, gives the same value, and the first and third levels are adjusted before the final factorial effects chart is drawn.

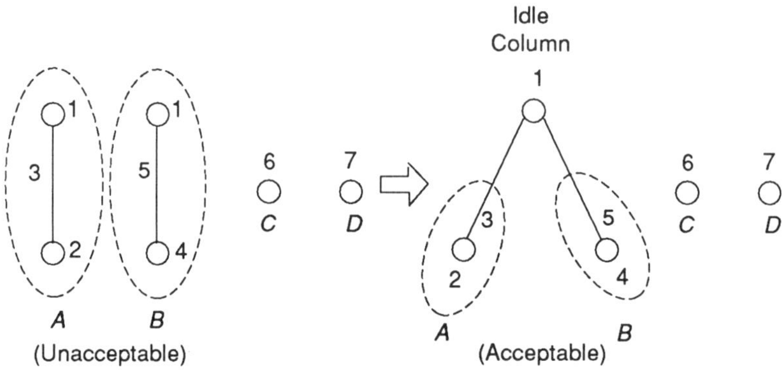

Figure 7-4: Assignment of Three Level Factors A and B and Two Level Factors C and D to an L_8 Using the Idle Column Technique

Table 7-5: Assignment of Three-Level Factors A and B and Two-Level Factors C and D to an L_8, Using the Idle-Column Technique

Factor / Column / No.	Idle	A	B	C	D
	1	2 3	4 5	6	7
1	1	1	1	1	1
2	1	1	2	2	2
3	1	2	1	2	2
4	1	2	2	1	1
5	2	2	2	1	2
6	2	2	3	2	1
7	2	3	2	2	1
8	2	3	3	1	2

7-4. Two-Level Arrangement in the Three-Level System (Dummy-Level Technique)

In order to have a two-level factor in a three-level orthogonal array, levels one, two, or three in the column assigned with the two-level factor should be duplicated. Suppose that among four factors, *A*, *B*, *C*, and *D*, only factor *B* has two levels and the rest have three levels. The linear graph and assignment to an L_9 are shown in *Figure 7-5* and *Table 7-6*, respectively.

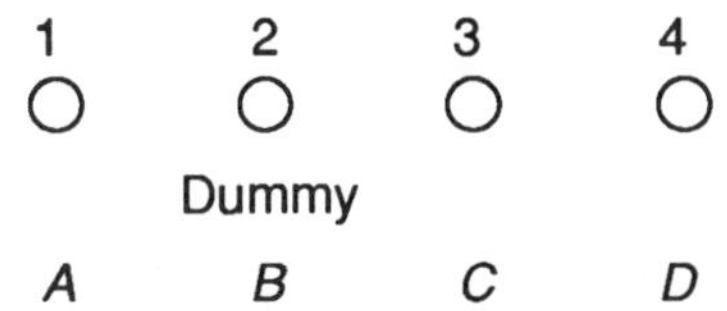

Figure 7-5: Linear Graph for the Assignment of Three Factors With Three Levels (*A*, *C*, *D*) and One Factor With Two Levels (*B*) to an L_9

Here, the levels of factor *B* are set as follows:

$$A_1 = A_1,\ A_2 = A_2,\ A_3 = A'_1$$

Table 7-6: An L_9 Orthogonal Array with Assignment of Three Factors with Three Levels (*A*, *C*, and *D*) and One Factor with Two Levels (*B*)

Factor / Column / No.	*A*	*B*	*C*	*D*
	1	2	3	4
1	1	1	1	1
2	1	2	2	2
3	1	1'	3	3
4	2	1	2	3
5	2	2	3	1
6	2	1'	1	2
7	3	1	3	2
8	3	2	1	3
9	3	1'	2	1

7-5. Nine-Level Arrangement in the Three-Level System

A nine-level factor can be arranged by using one line of the linear graph of a three-level system. For an L_{27}, this will be the following:

If columns 1 and 2 are used, the interaction columns will be columns 3 and 4. One line takes columns 1, 2, 3, and 4. The level combinations of the two end columns, i.e., (1, 1), (1, 2), (1, 3), (2, 1), (2, 2), (2, 3), (3, 1), (3, 2), and (3, 3), are replaced with 1, 2, 3, 4, 5, 6, 7, 8, and 9, and columns 1, 2, 3, and 4 are eliminated to complete one column with nine levels.

7-6. Combination Technique

This method treats two factors as one factor. This reduces the number of sacrificed columns and makes it possible to assign many factors within the same orthogonal array. Suppose that a spray experiment was carried out to improve washing by setting levels of pressure (P) and water quantity (Q) as shown in *Table 7-7*. It is not necessary to test all combinations of levels. It is only necessary to have levels of maximum strength, minimum strength, or medium strength of combined P and Q. If there is no difference among levels of (PQ), neither factor P nor Q is effective. If the combined factors have some effect, the effect of each factor can be checked separately.

Table 7-7: How To Make Combined Levels

P \ Q	3	5	7
2	$(PQ)_1$		
4		$(PQ)_2$	
6			$(PQ)_3$

(P:kg/cm^2) (Q:l/min)

7-7. A Method Where Different Factor Types Are Used for Different Levels (Nested-Factor Technique)

The nested-factor technique can be used when different types of factors are used for different levels. This technique is also known as pseudo-factor design.

Suppose that the alumite method (H_1) and the chemical chrome-plating method (H_2) are suggested for the undercoating of an aluminium paint plate. When heterogeneous levels are used, as in this case, the level assignment becomes complicated. For example, current density (D'_1, D'_2) and electrolyte temperature (T'_1, T'_2) are taken for H_1, and catalyst density (D''_1, D''_2) and plating temperature (T''_1, T''_2) are taken for H_2. Other common factors tested are paint (P_1, P_2) and heating temperature (B_1, B_2).

The only possible combination of H_1 is D' and T' and of H_2 is D'' and T''. D' and T'' should not be combined. The linear graph and assignment in this example are shown in *Figure 7-6* and *Table 7-8*, respectively.

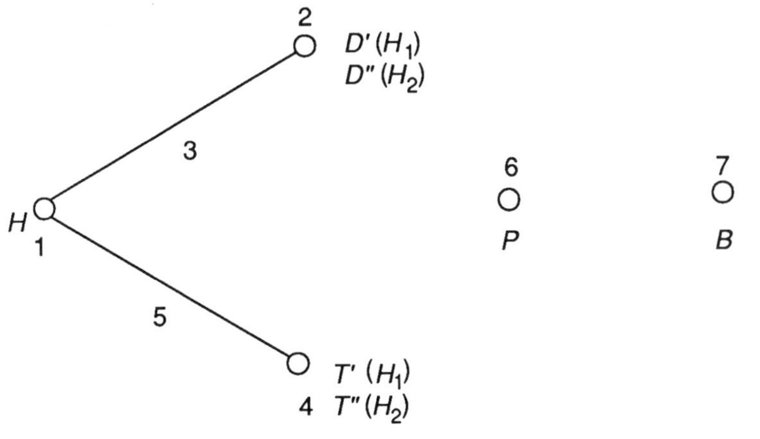

Figure 7-6: Assignment Based on the Nested-Factor Technique (L_8)

Table 7-8: Assignment Based On the Nested-Factor Technique

Factor	H	D' / D''		T' / T''		P	B
Column / No.	1	2 3		4 5		6	7
1	1	1	D'	1	T'	1	1
2	1	1		2		2	2
3	1	2		1		2	2
4	1	2		2		1	1
5	2	1	D''	1	T''	1	2
6	2	1		2		2	1
7	2	2		1		2	1
8	2	2		2		1	2

This kind of assignment, where the level combination changes from D' and T' for H_1 to D'' and T'' for H_2, is called the **nested-factor technique**. The optimum conditions are derived for both H_1 and H_2, and the final determination between the two is based on the comparison of the estimates.

Various assignment techniques together with the corresponding orthogonal arrays are summarized in *Table 7-9*. Among them, the L_{12}, L_{18}, and L_{36} are simple and easy to use, even for beginners, because they do not require complicated assignment techniques. I strongly recommend them.

Table 7-9: Major Assignment Techniques and Orthogonal Arrays

No.	Assignment Technique	Without-Interaction Column			With-Interaction Column				
		L_{12}	L_{18}*	L_{36}	L_8	L_{16}	L_{32}	L_9	L_{27}
1	4-level		O		O	O	O		O
2	8-level					O	O		O
3	3-level (dummy-level technique)				O	O	O		
4	3-level (idle-column technique)				O	O	O		
5	2-level (dummy-level technique)		O	O				O	O
6	9-level								O
7	Combination technique	O	O	O	O	O	O	O	O
8	Nested-factor technique		O		O	O	O		O

* For an L_{18}, the interaction between col. 1 and col. 2 can be obtained. Moreover, a 6-level column can be created using col. 1 and col. 2.

7-8. Assignment and Analysis Examples

This section uses examples of actual applications to illustrate various assignment and analysis techniques for deriving the optimum condition.

7-8-1. Example 1: Printed Circuit Board Test (L_{16}, Four Levels, Dummy-Level Technique, Idle-Column Technique)

This example relates to the production of a printed circuit board and concerns the chemical copper-plating process (which is one of approximately 10 processes required to produce a printed circuit board). It concerns the adhesiveness between the plated film and the baseplate. In order to improve the adhesiveness, the optimum condition of the preprocessing and chemical copper plating was derived.

Figure 7-7 shows the production process. *Table 7-10* shows the factors relating to etching, activation, and copper plating and their levels. Adhesiveness is measured by the peeling test results (Marbone standard). *Figure 7-8* and

Table 7-11 show the linear graph and assignment, respectively. Multi-level arrangement was done based on the idle-column technique for factors *A* and *B*, based on the standard multi-level arrangement for factor *C*, and based on the dummy-level technique for factor *D*.

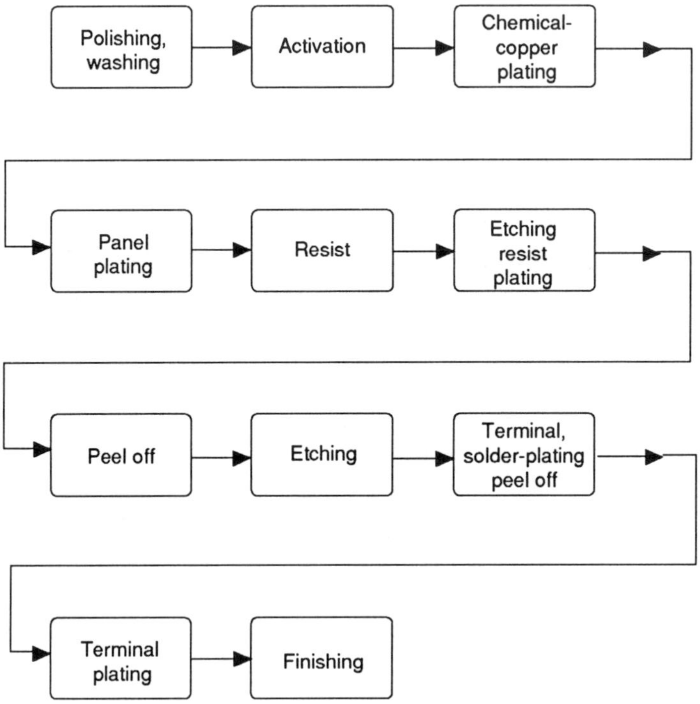

Figure 7-7: Production Process for a Printed Circuit Board

Table 7-10: Factors and Levels

Process	Factor	Level				
			1	2	3	4
Etching	Chromic sulfic acid temperature (°C)	*A*	55	60	65	
Activation	Palladium chloride density (g/l)	*B*	0.1	0.2	0.3	
Chemical-copper plating	Blue vitriol ($CuSO_4 \bullet 5\,H_2O$)	*C*	8	10	12	14
	Russel salt (g/l)	*D*	30	35	40	
	Sodium hydroxide (g/l)	*E*	8	10		
	Formalin (mg/l)	*F*	20	30		
	Temperature (°C)	*G*	18	24		
	Air stirring quantity (l/min)	*H*	10	20		

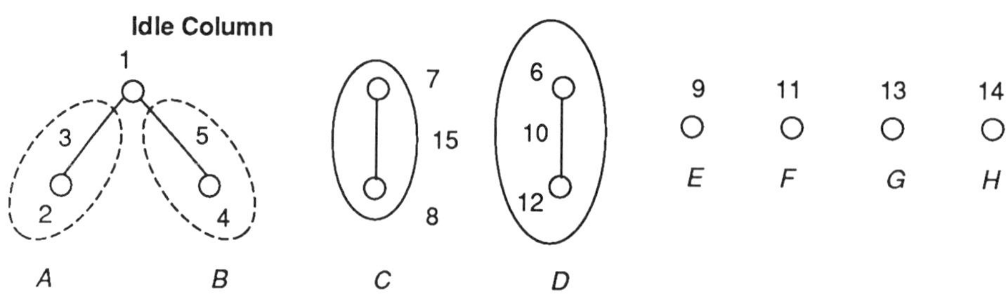

Figure 7-8: Linear Graph

Table 7-11: Assignment and Experimental Data

Column / No.	Idle	A	B	C	D	E	F	G	H	Data
	1	2, 3	4, 5	7, 8, 15	6, 10, 12	9	11	13	14	
1	1	1	1	1	1	1	1	1	1	44
2	1	1	1	2	2	2	2	2	2	33
3	1	1	2	3	2	1	1	2	2	29
4	1	1	2	4	3	2	2	1	1	41
5	1	2	1	3	3	1	2	1	2	48
6	1	2	1	4	2	2	1	2	1	31
7	1	2	2	1	2	1	2	2	1	28
8	1	2	2	2	1	2	1	1	2	39
9	2	2	2	3	1	2	2	2	1	29
10	2	2	2	4	2	1	1	1	2	33
11	2	2	3	1	2	2	2	1	2	43
12	2	2	3	2	3	1	1	2	1	30
13	2	3	2	1	3	2	1	2	2	22
14	2	3	2	2	2	1	2	1	1	38
15	2	3	3	3	2	2	1	1	1	37
16	2	3	3	4	1	1	2	2	2	34

1. Assignment

Four levels are used for factors *A*, *B*, *C*, and *D*. If the dummy-level technique is used to create three levels of *A*, *B*, and *D*, there will be a shortage of one column. The use of an L_{32} is possible, but the number of experiments will be impractically large. Thus, the idle-column technique is used because it allows three-level assignment without many columns.

On the other hand, if the dummy-level technique is used for all factors *A*, *B*, and *D* to create three levels, one column will remain unused. If this unused column is assigned with another two-level factor, all columns will be used. However, another factor is not assigned here, and the idle-column technique is applied only to factors *A* and *B*.

If column 1 is used as the idle column and columns 2 and 4 are assigned with factors *A* and *B*, columns 3 and 5 will be the interaction columns. The third levels of factors *A* and *B* are created by keeping the levels of *A* (column 2) and *B* (column 4) corresponding to level one of the idle column and by reading levels one and two of *A* (column 2) and *B* (column 4) corresponding to level two of the junction column as 2 and 3. As for factor *C*, the level combinations of columns 7 and 8, i.e., (1, 1), (1, 2), (2, 1), and (2, 2) are replaced with 1, 2, 3, and 4. For factor *D*, the level combination of columns 6 and 12, i.e., (1, 1), (1, 2), (2, 1), and (2, 2), are replaced with 1, 2, 3, and 2'. This completes the assignment. The assignment results and experimental data are shown in *Table 7-11*.

2. Data Analysis

Data for the adhesiveness between the base plate and the chemical-plated film were collected. Because adhesiveness is a Larger-the-Better (LTB) characteristic, it can be converted to an S/N ratio, but the observed data themselves (raw data) are used for the analysis.

First, the level sum is calculated. The level sum of factors *A* and *B* is calculated separately for levels one and two of the idle column.

Level one of the idle column: Idle (1)

- Level sum of A_1 $\quad A_1 = 44 + 33 + 29 + 41 = 147$
  Level sum of A_2 $\quad A_2 = 48 + 31 + 28 + 39 = 146$
  Level sum of B_1 $\quad B_1 = 44 + 33 + 38 + 31 = 156$
  Level sum of B_2 $\quad B_2 = 29 + 41 + 28 + 39 = 137$

  Level two of the idle column: Idle (2)

  Level sum of A_2 $\quad A_2 = 29 + 33 + 43 + 30 = 135$
  Level sum of A_3 $\quad A_3 = 22 + 38 + 37 + 34 = 131$
  Level sum of B_2 $\quad B_2 = 29 + 33 + 22 + 38 = 122$
  Level sum of B_3 $\quad B_3 = 43 + 30 + 37 + 34 = 144$
- Level sum of factor *C*
  Level sum of C_1 $\quad C_1 = 44 + 28 + 43 + 22 = 137$
  Level sum of C_2 $\quad C_2 = 33 + 39 + 30 + 38 = 140$
  Level sum of C_3 $\quad C_3 = 29 + 48 + 29 + 37 = 143$
  Level sum of C_4 $\quad C_4 = 41 + 31 + 33 + 34 = 139$
- Level sum of factor *D*
  Level sum of D_1 $\quad D_1 = 44 + 39 + 29 + 34 = 146$
  Level sum of D_2 $\quad D_2 = 33 + 28 + 33 + 37 = 131$
  Level sum of D_3 $\quad D_3 = 29 + 31 + 43 + 38 = 141$
  Level sum of D_4 $\quad D_4 = 41 + 48 + 30 + 22 = 141$
- Level sum of *E*
  Level sum of E_1 $\quad E_1 = 44 + 29 + 48 + 28 + 33 + 30 + 38 + 34 = 284$
  Level sum of E_2 $\quad E_2 = 33 + 41 + 31 + 39 + 29 + 43 + 22 + 37 = 275$
- Level sum of factors *F*, *G*, and *H*
  $F_1 = 265$ $\quad F_2 = 294$
  $G_1 = 323$ $\quad G_2 = 236$
  $H_1 = 278$ $\quad H_2 = 281$

  The results are summarized in *Table 7-12*.

Table 7-12: Level Sum

Level	Idle	*A*		*B*		*C*	*D*	*E*	*F*	*G*	*H*
		J(1)	*J*(2)	*J*(1)	*J*(2)						
1	293	147		156		137	146	284	265	323	278
2	266	146	135	137	122	140	131, 141	275	294	236	281
3	—		131		144	143	141				
4	—					139					

($T = 569$)

Next, the average is calculated. The average is calculated by dividing the level sum by the number of data added to derive the level sum. However, the mean of levels one and two of factors *A* and *B*, which are assigned based on the idle column technique needs to be modified so that level two, which is the common (reference) level, has the same average.

- Average of factor *A*

$$\text{Idle (1)} \begin{cases} \overline{A}_1 = \dfrac{147}{4} = 36.75 & \rightarrow 35.375 \\ \overline{A}_2 = \dfrac{146}{4} = 36.50 & \end{cases}$$

$$\rangle \qquad \rightarrow 35.125$$

$$\text{Idle (2)} \begin{cases} \overline{A}_2 = \dfrac{135}{4} = 33.75 & \\ \overline{A}_3 = \dfrac{131}{4} = 32.75 & \rightarrow 34.125 \end{cases}$$

The modified average of *A* is obtained by subtracting 1.375 from $\overline{A}_1$ and adding 1.375 to $\overline{A}_3$.

- Average of factor *B*

$$\text{Idle (1)} \begin{cases} \overline{B}_1 = \dfrac{156}{4} = 39.00 & \rightarrow 37.125 \\ \overline{B}_2 = \dfrac{137}{4} = 34.25 & \end{cases}$$

$$\rangle \qquad \rightarrow 32.375$$

$$\text{Idle (2)} \begin{cases} \overline{B}_2 = \dfrac{122}{4} = 30.50 & \\ \overline{B}_3 = \dfrac{14}{4} = 36.00 & \rightarrow 37.875 \end{cases}$$

The modified average of *A* is obtained by subtracting 1.875 from B_1 and adding 1.875 to B_3.

- Average of factor *C*

$$\overline{C}_1 = \frac{137}{4} = 34.25$$

$$\overline{C}_2 = \frac{140}{4} = 35.00$$

$$\overline{C}_3 = \frac{143}{4} = 35.75$$

$$\overline{C}_4 = \frac{139}{4} = 34.75$$

- Average of factor *D*

$$\overline{D}_1 = \frac{146}{4} = 36.50$$

$$\overline{D}_2 = \frac{(131+141)}{8} = 34.00$$

$$\overline{D}_3 = \frac{141}{4} = 35.25$$

- Average of factor *E*

$$\overline{E}_1 = \frac{284}{8} = 35.500$$

$$\overline{E}_2 = \frac{275}{8} = 34.375$$

The averages of other factors are calculated similarly:

$$\overline{F}_1 = 33.125 \qquad \overline{F}_2 = 36.750$$

$$\overline{G}_1 = 40.375 \qquad \overline{G}_2 = 29.500$$

$$\overline{H} = 34.750 \qquad \overline{H}_2 = 35.125$$

These calculation results are summarized in *Table 7-13*. The bracketed numbers indicate the average after modification.

Table 7-13: Process Average

Level	Idle	*A*		*B*		*C*	*D*	*E*	*F*	*G*	*H*
		J(1)	*J*(2)	J(1)	*J*(2)						
1	36.625	36.75 (35.375)		39.00 (37.125)		34.25	36.50	35.500	33.125	40.375	34.750
2	33.250	36.75 (37.125)	33.75 (35.125)	34.25 (32.375)	30.50 (32.375)	35.00	34.00	34.375	36.75	29.50	35.125
3			32.75 (34.125)		35.00 (37.875)	35.75	35.25				
4						34.75					

$$\overline{T} = \frac{559}{16} = 34.9375$$

The factorial effects chart, *Figure 7-9*, was drawn based on these averages.

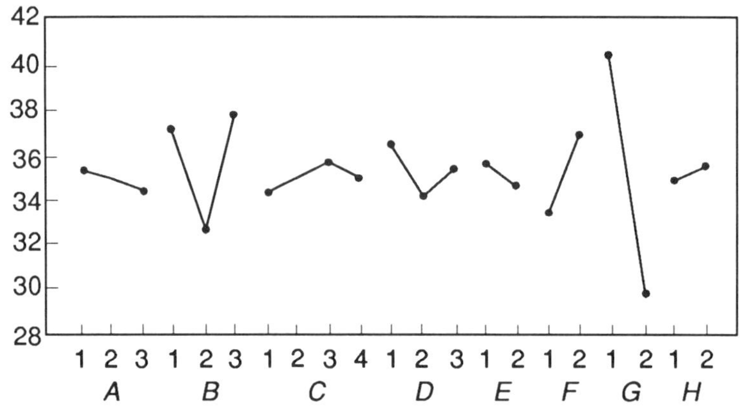

Figure 7-9: Factorial Effects

3. Optimum Condition Estimation

According to the factorial effects chart, factors *B*, *D*, *F*, and *G* are effective, and their optimum levels are B_3, D_1, F_2, and G_1.

The factorial levels of *A* and *B* corresponding to levels 1 and 2 of the junction column, i.e., A_1A_2, A_2A_3, and B_1B_2, B_2B_3, show differences in adhesiveness. These differences are the differences in the process average of the idle column. Because level one of the idle column gives a larger value in this example, the second optimum-condition candidate was determined to be B_1, D_1, F_2, and G_1. Although factor *A* has almost no effect, level A_1 is chosen since it gives a better result in A_1A_2, which belongs to level one of the junction column's process average. The adhesiveness of the two optimum condition candidates was calculated as follows:

$$\hat{\mu}(B_3D_1F_2G_1) = \overline{B}_3 + \overline{D}_1 + \overline{F}_2 + \overline{G}_1 - 3 \times \overline{T}$$

$$= 37.875 + 36.50 + 36.75 + 40.375 - 3 \times 34.9375$$

$$= 46.6875$$

$$\hat{\mu}(B_1D_1F_2G_1) = \overline{B}_1 + \overline{D}_1 + \overline{F}_2 + \overline{G}_1 - (3 \times \overline{T})$$

$$= 37.125 + 36.50 + 36.75 + 40.375 - 3 \times 34.9375$$

$$= 45.9375$$

The choice between these two candidates was made following the confirmatory trials.

7-8-2 Example 2: Surface-Washing Test (L_9, Combination Technique, Dummy-Level Technique)

1. Assignment

This example relates to the surface washing of an electronic component part. The washing process has three-level factors *A* and *D* and two-level factors *B*, *C*, and *E*. An L_9 is used to limit the number of experiments. Because there are five factors, two-level factors *B* and *C* are combined and assigned to one column. Factor *E* is assigned according to the dummy-level technique by duplicating E_3 as E'_1. *Table 7-14* shows the factors and levels, and *Table 7-15* shows the assignment and results of the experiment.

Table 7-14: Factors and Levels

Symbol	Factor Name	Level 1	Level 2	Level 3
A	Washing angle	30	45	60
B	Nozzle pressure	4	6	
C	Nozzle flow quantity	2	3	
D	Water temperature	40	50	60
E	Additive quantity	10	20	
(*BC*)	Combination of *B* and *C*	B_1C_1	B_1C_2	B_2C_2

Table 7-15: Assignment and Experimental Data

Factor / Column / No.	*A* 1	(*BC*) 2	*D* 3	*E* 4	Quantity of Residual Dirt	Transformed Data [db]
1	1	1	1	1	9, 10, 11	–20.029
2	1	2	2	2	5, 7, 6	–15.643
3	1	3	3	1'	7, 5, 4	–14.771
4	2	1	2	1'	21, 25, 19	–26.773
5	2	2	3	1	14, 16, 18	–24.127
6	2	3	1	2	10, 9, 11	–20.029
7	3	1	3	2	8, 8, 11	–19.191
8	3	2	1	1'	9, 8, 6	–17.806
9	3	3	2	1	4, 5, 8	–15.441

2. Data Analysis

The characteristic value here is the quantity of residual dirt. The data are transformed into Smaller-the-Better (STB) S/N ratios:

No. 1 $\eta_1 = -10\log\frac{1}{3}\left(9^2 + 10^2 + 11^2\right) = -20.029\,[\text{db}]$

The remaining data are similarly transformed:

No. 2 $\eta_2 = -10\log\frac{1}{3}(5^2 + 7^2 + 6^2) = -15.643 \quad [\text{db}]$

$\vdots$

No. 9 $\eta_9 = -10\log\frac{1}{3}(4^2 + 5^2 + 8^2) = -15.441 \quad [\text{db}]$

The level sum and the average are calculated as follows:
Level sum and average of factor A:

$$A_1 = -20.029 - 15.643 - 14.771 = -50.443 \quad \overline{A}_1 = \frac{-50.443}{3} = -16.814$$

$$A_2 = -26.733 - 24.127 - 20.029 = -70.929 \quad \overline{A}_2 = \frac{-70.929}{3} = -23.643$$

$$A_3 = -19.191 - 17.806 - 15.441 = -52.438 \quad \overline{A}_3 = \frac{-52.438}{3} = -17.479$$

Level sum and average of BC:

$$(BC)_1 = -20.029 - 26.773 - 19.191 = -65.993$$

$$\overline{(BC)}_1 = \frac{-65.993}{3} = -21.998$$

$$(BC)_2 = -15.643 - 24.127 - 17.806 = -57.576$$

$$\overline{(BC)}_2 = \frac{-57.576}{3} = -19.192$$

$$(BC)_3 = -14.771 - 20.029 - 15.441 = -50.241$$

$$\overline{(BC)}_3 = \frac{-50.24}{3} = -16.747$$

Because BC gives the combination level, the separate effects of B and C are calculated as follows:

Level sum and average of B_1B_2 at C_2:

$$B_1(C_2) = -15.643 - 24.127 - 17.806 = -57.576$$

$$\overline{B}_1(C_2) = \frac{-57.576}{3} = -19.192$$

$$B_2(C_2) = -14.771 - 20.029 - 15.441 = -50.241$$

$$\overline{B}_2(C_2) = \frac{-50.241}{3} = -16.747$$

Level sum and average of C_1C_2 at B_1:

$$C_1(B_1) = -20.029 - 26.773 - 19.191 = -65.993$$

$$\overline{C}_1(B_1) = \frac{-65.993}{3} = -21.998$$

$$C_2(B_1) = -15.643 - 24.127 - 17.806 = -57.576$$

$$\overline{C}_2(B_1) = \frac{-57.576}{3} = -19.192$$

Level sum and average of factor D:

$$D_1 = -20.029 - 20.029 - 17.806 = -57.864 \quad \overline{D}_1 = \frac{-57.864}{3} = -19.288$$

$$D_2 = -15.643 - 26.773 - 15.441 = -57.857 \quad \overline{D}_2 = \frac{-57.857}{3} = -19.286$$

$$D_3 = -14.771 - 24.127 - 19.111 = -58.089 \quad \overline{D}_3 = \frac{-58.089}{3} = -19.363$$

Level sum and average of factor E:

$$E_1 = -20.029 - 14.771 - 26.773 - 24.127 - 17.806 - 15.44 = -118.947$$

$$\overline{E}_1 = \frac{-118.947}{6} = -19.8245$$

$$E_2 = -15.643 - 20.029 - 19.191 = -54.863 \quad \overline{E}_2 = \frac{-54.863}{3} = -18.288$$

Table 7-16 summarizes the level sums and the averages, and *Figure 7-10* shows the factorial effects.

Table 7-16: Level Sum and Average (S/N)

	Level Sums			Average (db)		
Factor	1	2	3	1	2	3
A	-50.442	-70.929	-52.438	-16.814	-23.643	-17.479
(BC)	-65.993	-57.576	-50.241	-21.998	-19.192	-16.747
D	-57.864	-57.857	-58.089	-19.288	-19.286	-19.363
E	-118.947	-54.863		-19.8245	-18.288	
$B(C_2)$	-57.576	-50.241		-19.192	-16.747	
$C(B_1)$	-65.993	-57.576		-21.998	-19.192	

$T = -173.81 \quad \overline{T} = -173.81/9 = -19.312$

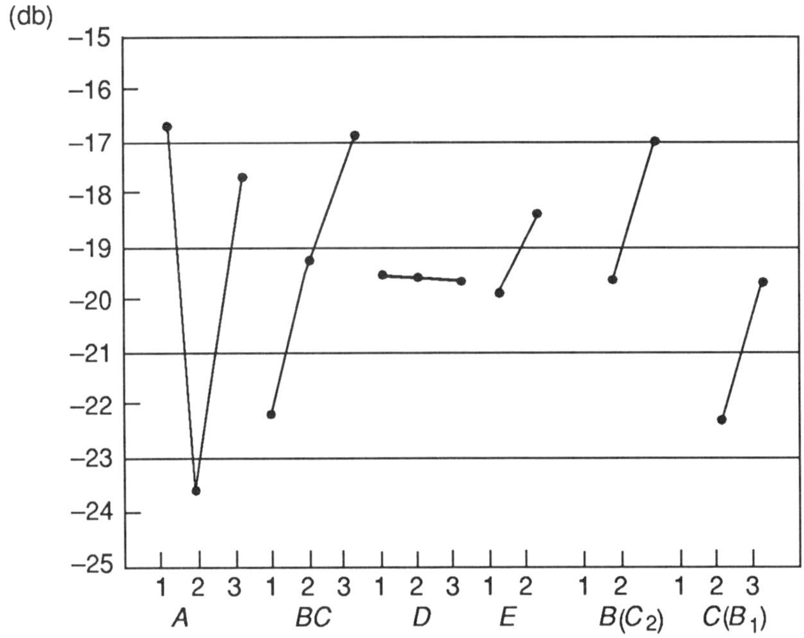

Figure 7-10: Factorial Effects

3. Optimum Condition Determination

According to the factorial effects chart, the effective factors are A, B, C, and D, and the optimum condition is the combination of $B_2C_2 = (BC)_3$ and either A_1 or A_3. The estimates are calculated for these conditions as follows:

$$\hat{\mu}(A_1B_2C_2) = \overline{A}_1 + \overline{(BC)}_3 - \overline{T} = -16.814 - 16.747 - (-19.312)$$

$$= -14.249 \quad [\text{db}]$$

$$\hat{\mu}(A_3B_2C_2) = \overline{A}_3 + \overline{(BC)}_3 - \overline{T} = -17.479 - 16.747 - (-19.312)$$

$$= -14.914 \quad [\text{db}]$$

The choice between these two candidates was made after the confirmatory trials by considering cost and operational ease. The estimates of the quantity of residual dirt were obtained through reverse log conversion as follows:

$$-14.249 = -10\log m^2 \qquad m = 5.16$$

$$-14.914 = -10\log m^2 \qquad m = 5.57$$

The estimates were at about 5.2 and 5.6, respectively.

7-8-3 Example 3: Metal-Surface Processing (L_{16}, Nested-Factor Technique)

This example relates to an aluminium printing plate. Surface processing of the printing plate is important in order to increase the number of copies made with one printing plate. A similar metal surface can be obtained by several different methods. The blast method and the electrolyte method are common methods of surface polishing. The control factors for the surface shape in these methods are different. The blast method utilizes mechanical factors, and the electrolyte method uses electrochemical factors. If the ordinary method of assignment is applied in this case, an unnatural combination may be created, e.g., an electrochemical factor corresponding to the blast method process. The experiment for the blast method should have only those factors that control the blast polish surface. For the factor polishing method, (C), the factors tested will be different for the blast method (level C_1) and for the electrolyte method (level C_2).

Factors *D* and *E* are taken in this example. They are shown in *Table 7-17*.

Table 7-17: Factors and Levels for Surface-Polishing Levels, According to the Nested-Factor Technique

C	Surface Polishing	*D*	Level 1	Level 2	*E*	Level 1	Level 2
C_1	Blast method	*D'*	Artificial alumina	Glass beads	*E'*	Vibration 30/min.	Vibration 60/min.
C_2	Electrolyte	*D''*	100 Clone	200 Clone	*E''*	Rectangular wave	Single-phase AC

The nested-factor method is applied to assign factors *C*, *D*, and *E* . The assignment is shown in *Figure 7-11*.

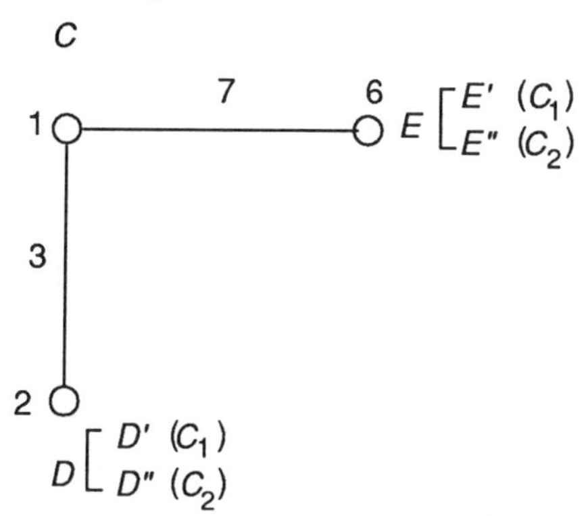

Figure 7-11 Nested-Factor Technique Linear Graph

Factors *C*, *D*, and *E* are assigned to columns 1, 2, and 6, respectively. The interaction appears in columns 3 and 7. The effects of *D* (*D′*, *D″*) and *E* (*E′*, *E″*) are obtained from columns 2 and 6.

Columns 3 and 7 are left unassigned. The linear graph for this part is shown in *Figure 7-11*, and the factors and levels are shown in *Table 7-17*.

With a different electrolyte (factor *J*), anodization needs a different kind of additive (factor *K*) to harden the oxide film and the material (factor *L*) of the layer between the under-layer (coated film) and oxide film. Factors *J*, *K*, and *L* are assigned. The linear graph for this portion is shown in *Figure 7-12* and the factors and levels are shown in *Table 7-18*.

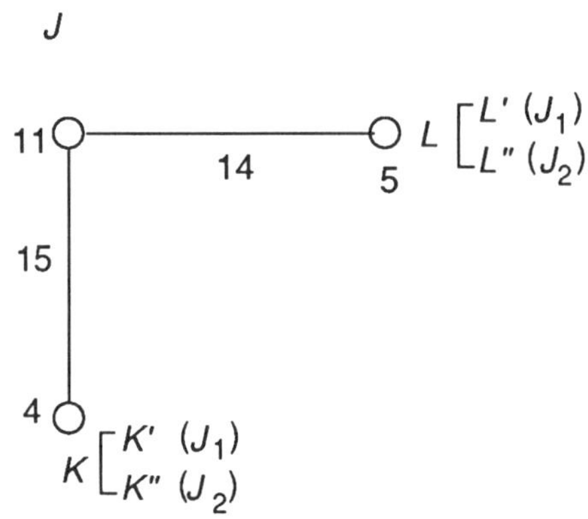

Figure 7-12: Nested-Factor Technique Linear Graph

Table 7-18: Factors and Levels for Anodization Levels (J_1 and J_2), According to the Nested-Factor Technique

J	Anodization	*K*	Level 1	Level 2	*L*	Level 1	Level 2
J_1	Sulfuric acid	*K′*	Without glycol	Glycol 5 g/l	*L′*	Poly-acrylic acid	Saponin
J''	Oxalic acid	*K″*	Without potassium oxalate	Potassium oxalate	*L″*	Poly-carboxylic acid	Poly-ethylene glycol

Factors *J*, *K*, and *L* are assigned to columns 11, 4, and 5, respectively, and the interaction appears in columns 15 and 14. No factor is assigned to columns 15 and 14.

Whereas factor sets *C*, *D*, and *E*, and *J*, *K*, and *L* are assigned according to the nested-factor technique, the rest of the (two-level) factors, i.e., *A*, *B*, *F*, *G*, and *H*, are assigned normally. The assignment is summarized in *Table 7-19*. A linear graph for all of the factors is shown in *Figure 7-13*.

Table 7-19: Factors and Levels

	Factor	Level 1	Level 2
A	Oxide film thickness (*m*)	0.2	0.4
B	Degreasing temperature interval (sec.)	30	60
F	Light sensitive film thickness (*m*)	0.8	1.0
G	Drying temperature (°C)	60	80
H	Developing time (sec.)	30	60

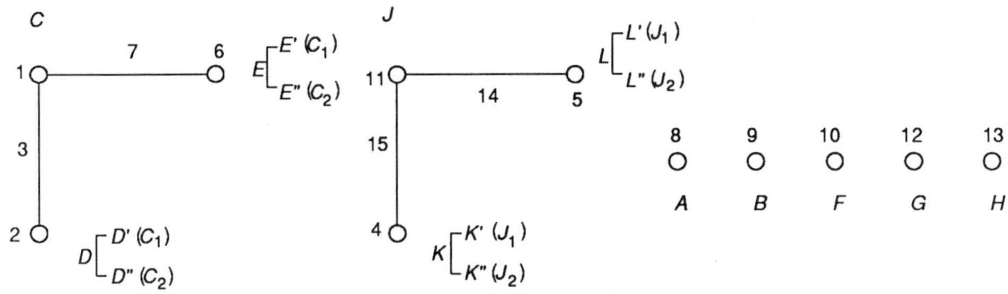

Figure 7-13: Assignment Linear Graph

The production process of the aluminium printing plate is shown in the following flow diagram (Figure 7-14).

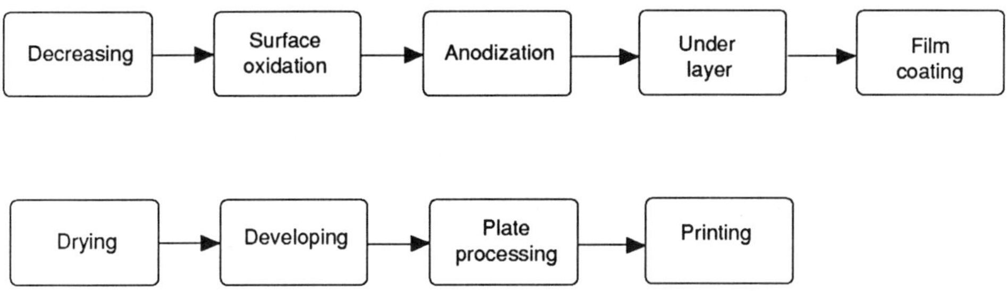

Figure 7-14: Flow Diagram for Aluminum Printing Plate Process

Assignment to the orthogonal array and the experimental results are shown in Table 7-20. *D′* and *E′* and *D″* and *E″* correspond to C_1 and C_2, respectively; *K′* and *L′* and *K″* and *L″* correspond to J_1 and J_2, respectively, although the assignment of factors *K* and *J* is complicated because factor *J* corresponds to column 11. This assignment makes it possible to obtain the optimum condition for each level of factors *C* and *J* from the same experimental samples, because suitable factors correspond to levels of factors *C* and *J*.

Table 7-20: Assignment and Results

Factor	C	D		E		J	K	L	A	B	F	G	H	No. of Copies
Column No.	1	2		6		11	4	5	8	9	10	12	13	Data
1	1	1′		1′		1	1′	1′	1	1	1	1	1	47
2	1	1′		1′		2	1″	1″	2	2	2	2	2	102
3	1	1′		2′		1	2′	2′	1	1	1	2	2	42
4	1	1′		2′		2	2″	2″	2	2	2	1	1	88
5	1	2′	D′	2′	E′	2	1″	1″	1	1	2	1	1	39
6	1	2′		2′		1	1′	1′	2	2	1	2	2	75
7	1	2′		1′		2	2″	2″	1	1	2	2	2	38
8	1	2′		1′		1	2′	2′	2	2	1	1	1	59
9	2	1″		1″		2	1″	2″	1	2	1	1	2	57
10	2	1″		1″		1	1′	2′	2	1	2	2	1	105
11	2	1″		2″		2	2″	1″	1	2	1	2	1	49
12	2	1″		2″		1	2′	1′	2	1	2	1	2	89
13	2	2″	D″	2″	E″	1	1′	2′	1	2	2	1	2	49
14	2	2″		2″		2	1″	2″	2	1	1	2	1	84
15	2	2″		1″		1	2′	1′	1	2	2	2	1	64
16	2	2″		1″		2	2″	1″	2	1	1	1	2	56

2. Data analysis

The experimental results are given as the number of copies. This is LTB data. However, the analysis here is based on the raw data.

The level sum and the process average are calculated for each factor:

$$C_1 = 47 + 102 + 42 + 88 + 39 - 75 + 38 + 59 = 490$$

$$\overline{C}_1 = \frac{490}{8} = 61.25$$

$$C_2 = 57 + 105 + 49 + 89 + 49 + 84 + 64 + 56 = 553$$

$$\overline{C}_2 = \frac{553}{8} = 69.125$$

$$D'_1 = 47 + 102 + 42 + 88 = 279 \qquad \overline{D}'_1 = \frac{279}{4} = 69.75$$

$$D'_2 = 39 + 75 + 38 + 59 = 211 \qquad \overline{D}'_2 = \frac{211}{4} = 52.75$$

$$D''_1 = 57 + 105 + 49 + 89 = 300 \qquad \overline{D}''_1 = \frac{300}{4} = 75$$

$$D''_2 = 49 + 84 + 64 + 56 = 253 \qquad \overline{D}''_2 = \frac{253}{4} = 63.25$$

$E'_1 = 47 + 102 + 38 + 59 = 246 \qquad \bar{E}'_1 = \frac{246}{4} = 61.5$

$E'_2 = 42 + 88 + 39 + 75 = 244 \qquad \bar{E}'_2 = \frac{244}{4} = 61.0$

$E''_1 = 57 + 105 + 64 + 56 = 282 \qquad \bar{E}''_1 = \frac{282}{4} = 70.5$

$E''_2 = 49 + 89 + 49 + 84 = 271 \qquad \bar{E}''_2 = \frac{271}{4} = 67.75$

$J_1 = 47 + 42 + 75 + 59 + 105 + 89 + 49 + 64 = 530$

$\bar{J}_1 = \frac{530}{8} = 66.25$

$J_2 = 102 + 88 + 39 + 38 + 57 + 49 + 84 + 56 = 513$

$\bar{J}_2 = \frac{513}{8} = 64.125$

$K'_1 = 47 + 75 + 105 + 49 = 276 \qquad \bar{K}'_1 = \frac{276}{4} = 69$

$K'_2 = 42 + 59 + 89 + 64 = 254 \qquad \bar{K}'_2 = \frac{254}{4} = 63.5$

$K''_1 = 102 + 39 + 57 + 84 = 282 \qquad \bar{K}''_1 = \frac{282}{4} = 70.5$

$K''_2 = 88 + 38 + 49 + 56 = 231 \qquad \bar{K}''_2 = \frac{231}{4} = 57.75$

$L'_1 = 47 + 75 + 89 + 64 = 275 \qquad \bar{L}'_1 = \frac{275}{4} = 68.75$

$L'_2 = 42 + 59 + 105 + 49 = 255 \qquad \bar{L}'_2 = \frac{255}{4} = 63.75$

$L''_1 = 102 + 39 + 49 + 56 = 246 \qquad \bar{L}''_1 = \frac{246}{4} = 61.5$

$L''_2 = 88 + 38 + 57 + 84 = 267 \qquad \bar{L}''_2 = \frac{267}{4} = 66.75$

$A_1 = 47 + 42 + 39 + 38 + 57 + 49 + 49 + 64 = 385$

$\bar{A}_1 = \frac{385}{8} = 48.125$

$A_2 = 102 + 88 + 75 + 59 + 105 + 89 + 84 + 56 = 658$

$\bar{A}_2 = \frac{658}{8} = 82.25$

The level sums and the means are similarly calculated for factors $B, F, G,$ and H. The calculation results are summarized in *Table 7-21*. *Figure 7-15* is the factorial effects chart.

Table 7-21: Level Sum and Average

Factor \ Level		Level Sum		Average	
		First	Second	First	Second
C		490	553	61.25	69.125
D	D'	279	211	69.75	52.75
	D''	300	253	75.00	63.25
E	E'	246	244	61.50	61.00
	E''	282	271	70.50	67.75
J		530	513	66.25	64.125
K	K'	276	254	69.00	63.50
	K''	282	231	70.50	57.75
L	L'	275	255	68.75	63.75
	L''	246	267	61.50	66.75
A		385	658	48.125	82.25
B		500	543	62.50	67.875
F		469	574	58.625	71.75
G		484	559	60.50	69.875
H		535	508	66.875	63.50

$T = 1043$ $\qquad \overline{T} = \frac{1043}{16} = 65.188$

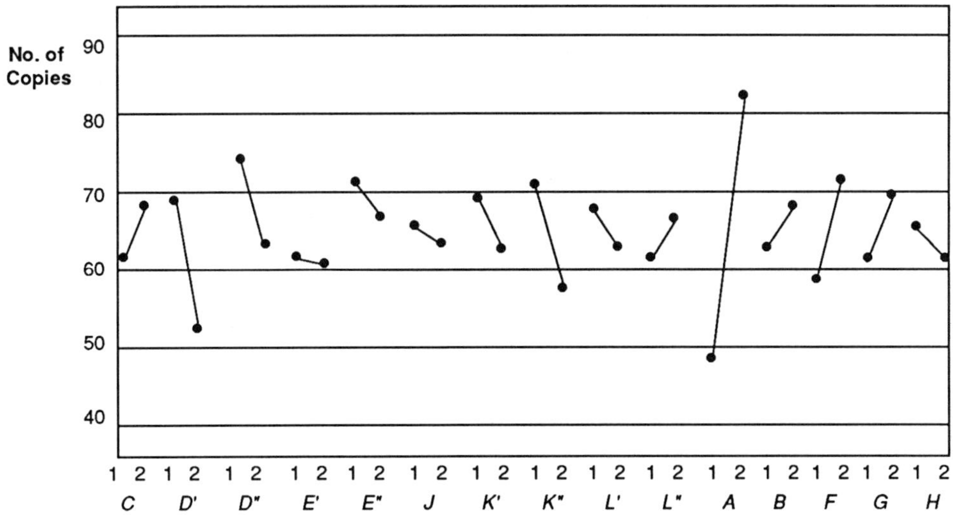

Figure 7-15: Factorial Effects Chart

3. Optimum condition estimate

First, the optimum condition is selected for each factor group created by the nested-factor technique.

- *C* Group

C_2 is the best level for factor *C*, and among the levels of C_2, D''_1 is the best. Factor *E* (E', E'') has almost no effect. The estimate of the combination of C_2 and D''_1 is as follows:

$$\hat{\mu}(C_2D''_1) = \overline{C}_2 + (\overline{D}''_1 - \overline{C}_2) = \overline{D}''_1 = 75$$

- *J* Group

Although Factor *J* did not give much difference, the best combinations are J_1, K'_1 and J_2, K''_1. Estimation is done for these two combinations:

$$\hat{\mu}(J_1K'_1) = \overline{J}_1 + (\overline{K}'_1 - \overline{J}_1) = \overline{K}'_1 = 69$$

$$\hat{\mu}(J_2K''_1) = \overline{J}_2 + (\overline{K}''_1 - \overline{J}_2) = \overline{K}''_1 = 70.5$$

Among other factors, *A*, *F*, and *G* are effective and their best levels are A_2, F_2, and G_2. The estimate of the total optimum condition is as follows:

$$\hat{\mu}(C_2D''_1\,J_1K'_1\,A_2F_2G_2) = \overline{D}''_1 + \overline{K}'_1 + \overline{A}_2 + \overline{F}_2 + \overline{G}_2 - 4 \times \overline{T}$$

$$= 75 + 69 + 82.25 + 71.75 + 69.875 - 4 \times 65.188$$

$$= 107.123$$

$$\hat{\mu}(C_2 D''_1 J_2 K''_1 A_2 F_2 G_2) = \overline{D}''_1 + \overline{K}''_1 + \overline{A}_2 + \overline{F}_2 + \overline{G}_2 - 4 \times \overline{T}$$

$$= 75 + 70.5 + 82.25 + 71.75 + 69.875 - 4 \times 65.188$$

$$= 108.623$$

These estimates are roughly the same, but condition C_2, D''_1, K'_1, A_2, F_2, G_2 is selected, because the anodization using J_1 (sulfuric acid) gives low cost and high productivity.

8

An Introduction to Analysis of Variance

8-1. The Role of Analysis of Variance

In the previous chapters, optimum levels were selected based on factorial effects on the characteristic values. These were expressed in terms of the line-slope patterns on the factorial effects chart. **Analysis of variance** (ANOVA) is a mathematical method that quantifies and verifies the relationship between factors and characteristic values that engineers judge intuitively.

According to analysis of variance, it is possible to say, for example, that the impact of increasing the quantity of an additive by two, three, and four times explains 30% of the changes of the entire experiment. It allows us to assess the degree to which a factor affects variation, rather than just saying that the additive has a significant effect. It is also possible to judge that the effect of an unknown factor (noise) is, say, 15%. The relative magnitude of the impact of factors on the characteristic values is expressed in terms of the variance, variance ratio, significance test, contribution ratio, etc. These are summarized in **ANOVA tables**. In short, analysis of variance is an instrument for mathematically quantifying a relationship between factors and characteristic values, which are obtained intuitively.

Analysis of variance is not an absolute necessity for experimental design. Rather, it is a supplement. That is why the majority of engineers in China and the United States do not learn analysis of variance for quality engineering. Although analysis of variance is not altogether unnecessary, it is not easy to master. Many engineers may abandon experimental design completely if their training course begins with it. In the United States, determination of the optimum condition in quality engineering, without analysis of variance, has been quite successful.

However, analysis of variance is required when applying tolerance design, Signal-to-Noise (S/N) ratios for dynamic characteristic analysis, measurement apparatus improvement, etc.

Furthermore, mastering analysis of variance allows one to obtain the maximum benefit from the experimental design method. This, however, should not be done at the beginning of training, because the difficulty in mastering this technique may make engineers abandon experimental design altogether.

This chapter explains the procedure for conducting analysis of variance. Complicated items such as variation, degrees of freedom, variance, and effective degrees of freedom calculations are summarized in *Table 8-5* through *Table 8-7*. After introducing analysis of variance, the examples from the previous chapters will be used to illustrate the application of the ANOVA method.

8-2. An Example of Analysis of Variance (L_8: Acid-Resistant Reactor Production)

This example relates to a steel reactor with an acid-resistant resin lining. The experiment was done in order to examine a method for improving adhesiveness of the lining film to the steel material.

8-2-1. Steps from Factor Listing to Average Calculation

The procedures for listing factors through average calculation are the same as those described in the previous chapters.

Many related factors are listed. The factors shown in *Figure 8-1* will be tested. The expected order of effectiveness against adhesiveness is *A*, *B*, *C*, *D*, *E*, and *F*. Because the interaction between curing-agent quantity (*A*) and drying temperature (*C*) cannot be anticipated, interaction information ($A \times C$) needs to be collected. *Table 8-1* summarizes the levels.

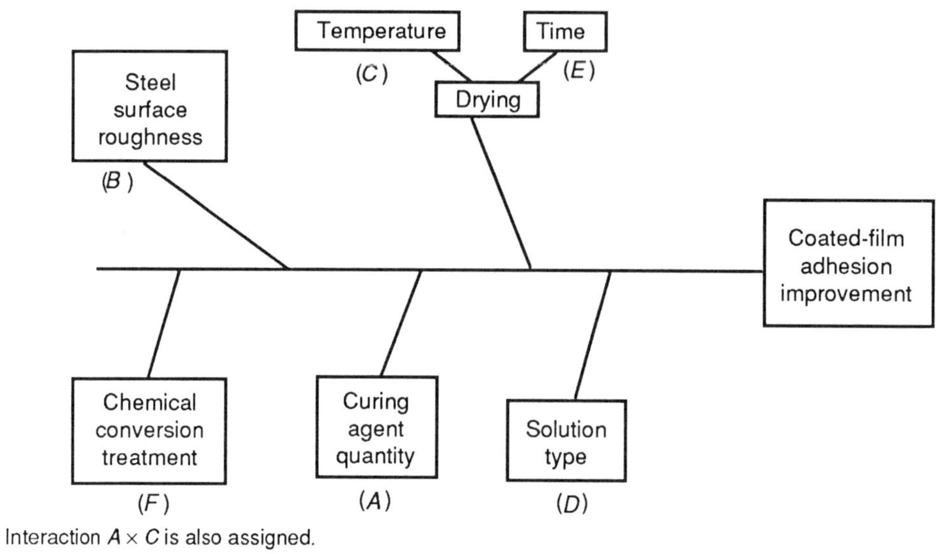

Figure 8-1: Factor Listing

Table 8-1: Level Determination

Symbol	Factor	Level	
		1	2
A	Curing agent quantity	2	4
B	Steel surface roughness	Rough	Fine
C	Temperature	180	160
D	Solution type	Toluene Series	Chlorine Series
E	Time	5	15
F	Chemical conversion treatment	No	Yes

It is necessary to have seven columns with two levels, including both the main effects and the interaction. Therefore, linear graph (1) of the L_8 is used. Column assignment is as follows:

1. Assign the main effects (factors whose individual effect should be tested) to the points — *A*, *B*, *C*, and *D*
2. Assign the interaction to the edge — $A \times C$
3. Assign the rest of main effects to the edges — *E*, *F*

Figure 8-2 shows the linear graph used for assignment. *Table 8-2* shows the experiment instructions where the factors are assigned to columns with actual levels in the order of experiments. This table also shows the experimental data.

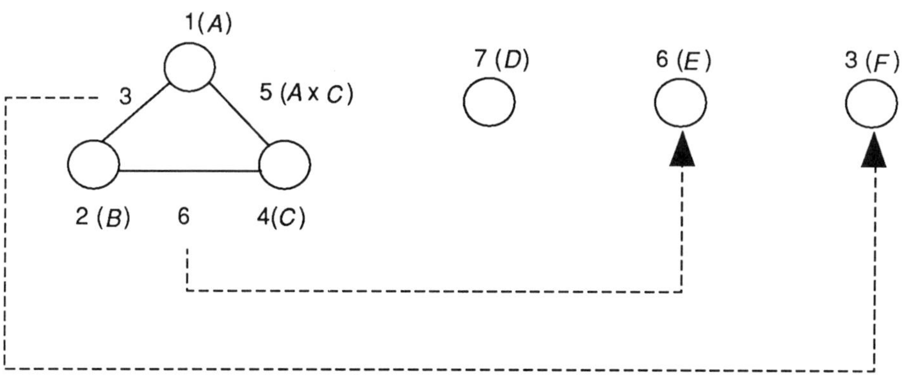

Figure 8-2: Linear Graph Used for Assignment

Table 8-2: Column Assignment, Experiment Instructions, and Experimental Data

Assignment Table $L_8 (2^7)$ Orthogonal Array								**Experiment Instructions and Data**								
Symbol	*A*	*B*	*F*	*C*	*A* × *C*	*E*	*D*	Experiment Instructions						Data		
Col. / No.	1	2	3	4	5	6	7	Steel Surface Roughness 2 (*B*)	Chemical Conversion Treatment 3 (*F*)	Curing Agent Quantity 1 (*A*)	Solution Type 7 (*D*)	Temp. 4(*C*)	Time 6 (*E*)	Acidity Resistance Time		Data Sum
1	1	1	1	1	1	1	1	Rough	No	2	Toluene	180	5	2	4	6
2	1	1	1	2	2	2	2	Rough	No	2	Chlorine	160	15	3	5	8
3	1	2	2	1	1	2	2	Fine	Yes	2	Chlorine	180	15	8	12	20
4	1	2	2	2	2	1	1	Fine	Yes	2	Toluene	160	5	6	4	10
5	2	1	2	1	2	1	2	Rough	Yes	4	Chlorine	180	5	10	7	17
6	2	1	2	2	1	2	1	Rough	Yes	4	Toluene	160	15	8	13	21
7	2	2	1	1	2	2	1	Fine	No	4	Toluene	180	15	7	10	17
8	2	2	1	2	1	1	2	Fine	No	4	Chlorine	160	5	4	1	5

Table 8-3: Level Sum and Average

(1) Main Effects

No.	Factor \ Level	Level Sum (1)	Level Sum (2)	Average (1)	Average (2)
1	A	44	60	5.5	7.5
2	B	52	52	6.5	6.5
3	F	36	68	4.5	8.5
4	C	60	44	7.5	5.5
5	$A \times C$	52	52	—	—
6	E	38	66	4.75	8.25
7	D	54	50	6.75	6.25
	(T)	(104)		(6.5)	

(2) Interaction

Level Sum

	C_1	C_2
A_1	26	18
A_2	34	26

Average

	C_1	C_2
A_1	6.5	4.5
A_2	8.5	6.5

Table 8-3 shows the level sum and the average. The level sum of the interaction between A and C is obtained by adding data corresponding to (A_1C_1), (A_1C_2), (A_2C_1), and (A_2C_2):

$$\text{Level sum} \quad A_1C_1 = (\text{Sum of data No.1 and3}) = (2+4)+(8+12) = 26$$

$$\text{Level sum} \quad A_1C_2 = (\text{Sum of data No.2 and4}) = (3+5)+(6+4) = 18$$

$$\text{Level sum} \quad A_2C_1 = (\text{Sum of data No.5 and7}) = (10+7)+(7+10) = 34$$

$$\text{Level sum} \quad A_2C_2 = (\text{Sum of data No.6 and8}) = (8+13)+(4+1) = 26$$

The average is calculated by dividing the level sum by the number of data, i.e., 4:

$$\overline{A_1C_1} = \frac{26}{4} = 6.5$$

$$\overline{A_1C_2} = \frac{18}{4} = 4.5$$

$$\overline{A_2C_1} = \frac{34}{4} = 8.5$$

$$\overline{A_2C_2} = \frac{26}{4} = 6.5$$

The factorial effects chart, *Figure 8-3*, is created based on the averages. Up to this point, the procedure for data analysis is the same as that used in the previous chapter. The next step in the analysis involves the calculations necessary for analysis of variance.

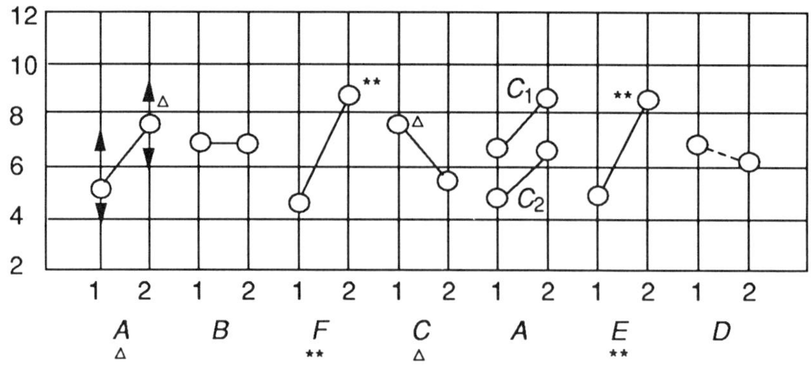

Figure 8-3: Factorial Effects

8-2-2. Variation Calculations and Degrees of Freedom

In order to conduct an analysis of variance, the variation and the degrees of freedom must to be calculated in order to create an analysis of variance table.

1. Calculate total variation, S_T

Total variation, S_T, is the numerical expression of the degree of variation of the entire experiment. It is given as the sum of squares of the difference between each datum (y_{ij}) and total mean ($\overline{T}$).

$$\text{Total variation}\quad S_T = (y_{11}-\overline{T})^2 + (y_{12}-\overline{T})^2 + \cdots + (y_{ij}-\overline{T})^2$$

$$= (y_{11}^2 + y_{12}^2 + \cdots + y_{ij}^2) - \frac{1}{i \times j}(y_{11} + y_{12} + \cdots + y_{ij})^2$$

where $y_{11} + y_{12} + \cdots + y_{ij} = T$. The given number of data is N. $CF = T^2/N$ is called the **correction term**. The total variation can then be written as:

S_T = sum of squares of each data point – CF

The total variation of our example experiment was calculated. The difference between each data point and the mean is given as:

$$\hat{a}_{ij} = \left(y_{ij} - \overline{T}\right)$$

The lengths of solid lines in *Figure 8-4* (1), which are the difference between each data point and the mean, are squared and added to obtain the **total variation** S_T. Because $\overline{T} = 104/16 = 6.5$, we have the following:

$$S_T = (2-6.5)^2 + (4-6.5)^2 + (3-6.5)^2 + \cdots + (1-6.5)^2$$

$$= (\hat{a}_{11})^2 + (\hat{a}_{12})^2 + (\hat{a}_{21})^2 + \cdots + (\hat{a}_{82})^2$$

$$= 2^2 + 4^2 + 3^2 + \cdots + 1^2 - (2+4+3+\cdots+1)^2/(2\times 8)$$

$$= 2^2 + 4^2 + 3^2 + \cdots + 1^2 - (104)^2/16 = 186$$

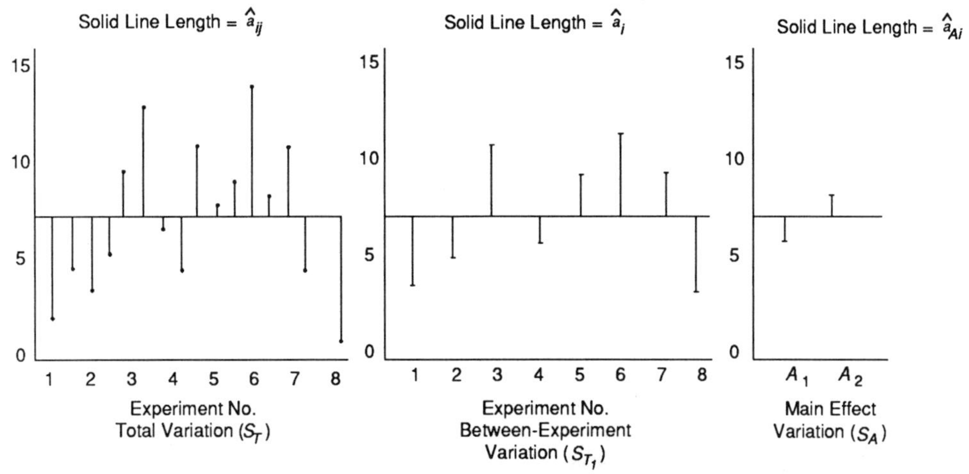

Figure 8-4: Concept of Total Variation S_T, Between Experiment Variation S_{T_1}, and Main Effect Variation S_A

Thus, the variation of the entire experiment is 186. The number of pieces of information is referred to as the **degrees of freedom**. The amount of information (degrees of freedom) is the same as the number of data. Because the total variation is found by squaring the sum of the differences from the total mean, whose number of information pieces is 1, the amount of information (degrees of freedom) is given by the number of data minus one. Letter f is used to indicate the degrees of freedom.

Degrees of freedom of total variation:

f_T = the number of data – 1 = 16 – 1 = 15

2. Calculate of between-experiment variation, S_{T_1}

Between-experiment variation, S_{T_1}, shows the extent of changes in the combination of each experiment in the orthogonal array. It is obtained by taking the squared differences ($\hat{a}_i$) between the mean ($\overline{y}_i$) of each experiment and the total mean ($\overline{T}$), and multiplying them by the number of replications before summing them up. The magnitude of $\hat{a}_i$ is shown in *Figure 8-4* (2).

The means of each experiment are:

$$\overline{y}_1 = \frac{(2+4)}{2} = 3,\quad \overline{y}_2 = \frac{(3+5)}{2} = 4, \cdots, \overline{y}_8 = \frac{(4+1)}{2} = 2.5$$

Between-experiment variation is:

$$S_{T_1} = 2(\overline{y}_1 - \overline{T})^2 + 2(\overline{y}_2 - \overline{T})^2 + \cdots + 2(\overline{y}_8 - \overline{T})^2$$

$$= 2(\hat{a}_1)^2 + 2(\hat{a}_2)^2 + \cdots + 2(\hat{a}_8)^2$$

If there are r repetitions, the above equation becomes:

$$S_{T_1} = \frac{1}{r}(Y_1^2 + Y_2^2 + \cdots + Y_8^2) - CF$$

where ΣY_i is the data sum.

Between-experiment variation in our example is calculated as follows:

$$S_{T_1} = 2\times(3-6.5)^2 + 2\times(4-6.5)^2 + \cdots + 2\times(2.5-6.5)^2$$

$$= \frac{1}{2}(6^2 + 8^2 + \cdots + 5^2) - \frac{(104)^2}{16}$$

$$= 146$$

Therefore, the extent of variation created by the experiment combination is 146. The number of degrees of freedom is $8 - 1 = 7$, because it is calculated from the mean of each experiment, $\overline{y}_i$, and the total mean, $\overline{T}$. In general,

Degrees of freedom of between-experiment variation is:

f_{T_1} = number of experiments in the orthogonal array - 1.

If there is no repetition, it is not necessary to calculate between-experiment variation.

3. Calculate repetition error variation, S_{e_2}

If the repeated data of each experiment are exactly the same, total variation (S_T) and between-experiment variation (S_{T_1}) will be the same. However, repeated data give different results each time, due to differences in level setting or the experimental environment (noise component). Therefore, the difference between total variation (S_T) and between experiment variation (S_{T_1}) is **repetition error variation** (S_{e_2}).

Repetition - error variation: $S_{e_2} = S_T - S_{T1}$

Degrees of freedom of repetition error variation: $f_{e_2} = f_T - f_{T1}$

Repetition error variation in our example and the degrees of freedom are obtained as follows:

$$S_{e_2} = 186 - 146 = 40$$

$$f_{e_2} = 15 - 7 = 8$$

Repetition error variation is also called **second order error variation.**

4. Calculate factor variation

Factor variation is a numerical expression of the impact of assigned factors on the characteristic value. It is calculated by taking the squared sum of the difference, $\hat{a}_{Ai}$, between average, $\overline{A}_i$, and total mean, $\overline{T}$, after multiplying the difference by the number of data used to calculate the average. The magnitude of $\hat{a}_{Ai}$ is indicated in *Figure 8-4*.

Factor A variation:
$$S_A = 8\times(\overline{A}_1-\overline{T})^2+8\times(\overline{A}_2-\overline{T})^2$$
$$= 8\times(\hat{a}_{A_1})^2+8\times(\hat{a}_{A_2})^2$$
$$= 8\times(5.5-6.5)^2+8\times(7.5-6.5)^2$$
$$= 8\times1+8\times1=16$$

The factor variation of other factors can be similarly calculated.

$$S_B = 8\times(6.5-6.5)^2+8\times(6.5-6.5)^2$$
$$= 0$$

$$S_F = 8\times(4.5-6.5)^2+8\times(8.5-6.5)^2$$
$$= 32+32=64$$

Because the calculation of factor variation from the process average data is complicated, the direct variation calculation formulae from the level sum data, shown in *Table 8-7*, can be used instead.

$$S_E = \frac{1}{8}(E_1^2-E_2^2)-\frac{T^2}{16}$$
$$= \frac{1}{16}(E_1-E_2)^2$$
$$= \frac{1}{16}(38-66)^2$$
$$= 49$$

$$S_D = \frac{1}{8}(D_1^2-D_2^2)-\frac{T^2}{16}$$
$$= \frac{1}{16}(D_1-D_2)^2$$
$$= 1$$

Factor variation is obtained as the difference between the average of each level and the total mean. Its degrees of freedom can be expressed in the following way:

Factor variation degrees of freedom = number of factor levels - 1.

The degrees of freedom for two-level factors ($A \sim F$) is 2 - 1 = 1. The degrees of freedom of factor A is written as f_A, and $f_A = 2 - 1 = 1$.

Variation of interaction $A \times C$ is given as:

$$S_{A\times C} = \frac{1}{4}(26^2+18^2+34^2+26^2)-\frac{(104)^2}{16}-S_A-S_C$$
$$= 32-16-16=0$$

The degrees of freedom for the interaction is the product of the degrees of freedom of interacting factors, i.e.,

$$f_{A\times C} = f_A \times f_C = 1\times 1 = 1$$

8-2-3. Formulae for Calculation of Multi-Level Factor Variation

Calculating the variation of a multi-level factor from the average and the total mean, $\overline{T}$, is complicated. A general formula should be used. **Variation, S_A, of a multi-level factor A** is obtained as follows:

$$\begin{aligned}
S_A &= n_1(\hat{a}_1)^2 + n_2(\hat{a}_2)^2 + n_3(\hat{a}_3)^2 + \cdots \\
&= n_1\left(\frac{A_1}{n_1} - \frac{T}{N}\right)^2 + n_2\left(\frac{A_2}{n_2} - \frac{T}{N}\right)^2 + n_3\left(\frac{A_3}{n_3} - \frac{T}{N}\right)^2 + \cdots \\
&= n_1\left\{\left(\frac{A_1}{n_1}\right)^2 - 2\left(\frac{A_1}{n_1}\right)\left(\frac{T}{N}\right) + \left(\frac{T}{N}\right)^2\right\} \\
&\quad + n_2\left\{\left(\frac{A_2}{n_2}\right)^2 - 2\left(\frac{A_2}{n_2}\right)\left(\frac{T}{N}\right) + \left(\frac{T}{N}\right)^2\right\} \\
&\quad + n_3\left\{\left(\frac{A_3}{n_3}\right)^2 - 2\left(\frac{A_3}{n_3}\right)\left(\frac{T}{N}\right) + \left(\frac{T}{N}\right)^2\right\} + \cdots \\
&= \frac{A_1^2}{n_1} + \frac{A_2^2}{n_2} + \frac{A_3^2}{n_3} + \cdots \\
&\quad -2(A_1 + A_2 + A_3 + \cdots) \times \left(\frac{T}{N}\right) \\
&\quad +(n_1 + n_2 + n_3 + \cdots) \times \left(\frac{T}{N}\right)^2 \\
&= \frac{A_1^2}{n_1} + \frac{A_2^2}{n_2} + \frac{A_3^2}{n_3} + \cdots - \frac{T^2}{N}
\end{aligned}$$

$$A_1 + A_2 + A_3 + \cdots = T, \quad n_1 + n_2 + n_3 + \cdots = N$$

where A_i is the level sum, n_i is the number of data used to calculate the level sum, T is the total, and N is the total number of data. The number of degrees of freedom is the number of levels minus one. These can be summarized as follows:

Variation Calculation Formula

$$\begin{bmatrix} S_A = \dfrac{A_1^2}{n_1} + \dfrac{A_2^2}{n_2} + \dfrac{A_3^2}{n_3} + \cdots - \dfrac{T^2}{N} \\ f_A = \text{number of levels} - 1 \end{bmatrix}$$

A_i: Level sum
n_i: The number of data used to calculate A_i
T: Total sum
N: Total number of data

Table 8-7 summarizes the variation calculation formulae for factors assigned to the orthogonal array based on the above formula. These formulae should be used carefully as supplementary formulae for calculating factor variation. I hope they will help readers master the experimental design method.

Table 8-4: Symbols Used in Analysis of Variance

Symbol	Content
L	Number of experiments in the orthogonal array
r	Number of experiment repetitions
T	Total sum of the data
N	Total number of data
CF	Correction factor
A_i	Level sum of i-th level of factor A
S_T	Total variation
S_{T_1}	Between-experiment variation
S_A	Factor A variation
$S_{A'}$	Factor A pure variation
S_{e_2}	Repetition-error variation
S_{e_1}	Between-experiment error variation
V_e	Error variance
V_{e_1}	Repetition error variance
V_{e_2}	Between-experiment error variance
f_T	Degrees of freedom of total variance
f_{T_1}	Degrees of freedom between experiment variation
f_A	Degrees of freedom of factor A
f_{e_2}	Degrees of freedom of repetition error
f_{e_1}	Degrees of freedom of between-experiment error
y_j	Data without repetition
y_{ij}	Data with repetition
Y_j	Sum of the data of each experiment on the orthogonal array

8-2-4. Construction of the ANOVA Table

Once degrees of freedom and variation are calculated, they are summarized in an ANOVA table. *Table 8-5* is an example of an analysis of variance table in which factor names, degrees of freedom, and variation are listed. At this stage, we should confirm that the sum of factor variation coincides with between-experiment variation, S_{T_1}.

$$\text{Sum of factor variation} = S_A + S_B + S_F + S_C + S_{A\times C} + S_E + S_D$$
$$= 16 + 0 + 64 + 16 + 0 + 49 + 1$$
$$= 146$$

The sum of factor variation should be the same as between-experiment variation. If they are not the same, there must be an error in calculation.

Table 8-5: ANOVA Table

Factor	*f* (Degrees of Freedom)	*S* (Variation)	*V* (Variance)	*Fo* (Variance Ratio)	Test	Percent Contribution ρ %
A	1	16	16	4.29	Δ	6.60
B	[1]	[0]	0 ✓	—	—	—
F	1	64	64	17.16	**	32.40
C	1	16	16	4.29	Δ	6.60
A × *C*	[1]	[0]	0 ✓	—	—	—
E	1	49	49	13.14	**	24.34
D	[1]	[1]	1 ✓	—	—	—
e_2	[8]	[40]	5 ✓			
T	15	186				100.00
e	[(11)]	[(41)]	(3.73)			(30.06)

1. Calculate factor variance and new error variance, V_e

   Variation, *S*, per degrees of freedom, *f*, is called variance. The magnitude of variance indicates the size of the factorial effect. **Variance**, *V*, is thus calculated by dividing variation, *S*, by the degrees of freedom, *f*:

$$\text{Factor variance} \quad V_A = \frac{S_A}{f_A} = \frac{16}{1} = 16$$

$$V_B = \frac{S_B}{f_B} = \frac{0}{1} = 0$$

The variance of other factors is similarly calculated:

$$V_F = 64, \quad V_C = 16, \quad V_{A\times C} = 0, \quad V_E = 49, \quad V_D = 1$$

Repetition error variance is given as:

$$V_{e_2} = \frac{S_{e_2}}{f_{e_2}} = \frac{40}{8} = 5$$

These variances are recorded in *Table 8-5*.

According to the table, factor variances V_B, $V_{A\times C}$, and V_D are smaller than second-order error variance V_{e_2}. Therefore, we can ignore the first order error, V_{e_1} (also called between-experiment error variance), and calculate a **new error variance,** V_e. The procedure of adding error to obtain a new error variance is referred to as the "**pooling of error.**"

The variation of factors with small variance, i.e., S_B, $S_{A\times C}$, and S_D, is pooled with S_{e_2} to obtain a new error variance, S_e:

$$S_e = S_B + S_{A\times C} + S_D + S_{e_2}$$

$$= 0 + 0 + 1 + 40$$

$$= 41$$

The **degrees of freedom for the new variance,** f_e, becomes:

$$f_e = f_B + f_{A\times C} + f_D + f_{e_2}$$

$$= 1 + 1 + 1 + 8$$

$$= 11$$

Therefore, the new error variance, V_e, is:

$$V_e = \frac{S_e}{f_e} = \frac{41}{11} = 3.73.$$

The degrees of freedom and variation of pooled factors are marked with a check (√).

If information exists on first order error variation, S_{e_1}, a **variance ratio** V_{e_1}/V_{e_2} should be calculated. If it is significant, new error variation is created within the between-experiment variation.

If there is no information about first order error variation, S_{e_1}, a new error component, V_e can be created by pooling the first and second order error variation, provided the smallest factor variance is less than about three times the second order error variance, V_{e_2}.

The degrees of freedom, variation, and variance of the new error term are listed in the bracket at the bottom of the ANOVA table. It is not necessary to obtain a variance ratio (F_0), significance test, pure variation (S'), or contribution ratio (ρ), for pooled factors. In our example, these are *B*, *A*×*C*, and *D*. The corresponding entry is marked with a bar (*Table 8-5*).

The number of factors whose variation can be pooled into the error variation should be about one-half to two-thirds of all assigned factors.

2. Calculate the variance ratio, F_0

The relative size of unpooled factor variance, *V*, against new error variance, V_e, is called the variance ratio, F_0. The **variance ratio** is obtained by dividing unpooled factor variance, *V*, by new error variance, V_e:

$$F_A = \frac{V_A}{V_e} = \frac{16}{3.73} = 4.29$$

$$F_F = \frac{V_F}{V_e} = \frac{64}{3.73} = 17.16$$

$$F_C = \frac{V_C}{V_e} = \frac{16}{3.73} = 4.29$$

$$F_E = \frac{V_E}{V_e} = \frac{49}{3.73} = 13.14$$

3. Significance Test

The **significance test** is a traditional statistical method. It is conducted by reading the F value from the F table corresponding to the degrees of freedom of the factor being considered (the degrees of freedom of the numerator in the F Table) and the degrees of freedom of new error variance, f_e (the degrees of freedom of the denominator in the F Table). Factors are marked with ** if the factor variance ratio is larger than the F value of the 1% level of significance, with * if at a level of significance between 1% and 5%, and with D if the level of significance is less than 5%. In our example, the F value is:

$$F_{11}^{1} = \begin{cases} 4.84 & (5\% \text{ level of significance}) \\ 9.65 & (1\% \text{ level of significance}) \end{cases}$$

Here factors are marked with ** if the variance ratio is greater than 9.65, with * if between 4.84 and 9.65, and with Δ if less than 4.84.

Those factors marked with ** and * are said to be significant at the 1% and 5% levels of significance, respectively. Marking factors F and E with ** and factors A and C with Δ completes the test.

4. Calculate contribution ratio, ρ

The **contribution ratio** is the percentage rate of the pure factorial effect, S', in the total variation, S_T. Factor variation, S, is the sum of pure variation, S', and the product of factor degrees of freedom and error variance:

$$S = S' + (f \times V_e)$$

Therefore, the factor contribution ratio, ρ, is given by the following:

$$\rho = \frac{\text{pure variation } (S')}{\text{total variation } (S_T)} \times 100 = \frac{S - (f \times V_e)}{S_T} \times 100 \ (\%)$$

The contribution ratio in our example is:

$$S_A' = S_A - f_A \times V_e = 16 - 1 \times 3.73 = 12.27$$

$$\rho_A = \frac{12.27}{186} \times 100 = 6.60 \ (\%)$$

$$S_F' = S_F - f_F \times V_e = 64 - 1 \times 3.73 = 60.27$$

$$\rho_F = \frac{60.27}{186} \times 100 = 32.40\,(\%)$$

$$S_C' = S_C - f_C \times V_e = 16 - 1 \times 3.73 = 12.27$$

$$\rho_C = \frac{12.27}{186} \times 100 = 6.60\,(\%)$$

$$S_E' = S_E - f_E \times V_e = 49 - 1 \times 3.73 = 45.27$$

$$\rho_E = \frac{45.27}{186} \times 100 = 24.34\,(\%)$$

The **contribution ratio of the error term** is obtained by subtracting the contribution ratios of unpooled factors from 100%:

$$\rho_e = 100 - (\rho_A + \rho_F + \rho_C + \rho_E)$$

$$= 100 - (6.60 + 32.40 + 6.60 + 24.34) = 100 - 69.94$$

$$= 30.06\,(\%)$$

Therefore, there is a 30.06% influence of unknown (noise) factors on the experimental results.

The preceding calculations complete ANOVA *Table 8-5*. In the table, the size of the effect of factors A - D are quantitatively expressed in terms of variance (V), variance ratio (F_0), significance test, and contribution ratio, (ρ). These quantified results coincide with the more intuitive results obtained from the factorial effects analysis.

8-2-5. Process average confidence limit

This section describes the determination of the **confidence limit of the process average,** which is obtained from the F Table as follows:

$$\text{Confidence limit} = \pm\sqrt{F_{fe}^{1}(0.05) \times \frac{V_e}{n_e}}$$

where V_e: New error variance

n_e: The effective number of replications

$F_{fe}^{1}\,(0.05)$: F value of 5% level of significance with degrees of freedom 1 and f_e

The **effective number of replications,** n_e, indicates how many replications were done to obtain the process average. It is eight for factors A, F, C, and E in our example. It will be four for interaction $A \times C$ if the interaction is significant. n_e can be obtained as follows:

Effective number of replications (n_e):

$$n_e = \frac{\text{Total number of experiments}}{1 + \text{Sum of degrees of freedom of unignored factors}}$$

Because other factors are ignored in the calculation of the process average, the effective number for replications of factor A is:

$$n_e = \frac{16}{1 + \text{Degree of freedom of factor } A \text{ (1)}} = \frac{16}{2} = 8$$

The confidence limit of the process average of factors $A, F, C,$ and E is given as follows, because the F value at 5% significance level with 1 degree for freedom for the numerator and 11 degrees of freedom for the denominator ($F_{11}^{1}(0.05)$) is 4.48, the error variance (V_e) is 3.37, and the effective number of replications, (n_e), is 8:

$$\text{Confidence limit} = \pm\sqrt{4.84 \times \frac{3.73}{8}} = \pm 1.50$$

This confidence limit should be drawn on the factorial effects chart. Showing the limits of factor A alone is sufficient.

8-2-6. The Optimum Condition and Its Confidence Limit

The main characteristic value in our example is adhesiveness. This is a Larger-the-Better (LTB) characteristic. Therefore, an estimate of the process average, $\hat{\mu}$, is calculated for optimum condition levels of significant factors A_2, F_2, C_1, and E_2:

$$\begin{aligned}\hat{\mu}(A_2F_2C_1E_2) &= \overline{T} + (\overline{A}_2 - \overline{T}) + (\overline{F}_2 - \overline{T}) + (\overline{C}_1 - \overline{T}) + (\overline{E}_2 - \overline{T}) \\ &= \overline{A}_2 + \overline{F}_2 + \overline{C}_1 + \overline{E}_2 - 3 \times \overline{T} \\ &= 7.5 + 8.5 + 7.5 + 8.25 - 3 \times \overline{T} \\ &= 12.25\end{aligned}$$

Because factors that should not be ignored are $A, F, C,$ and E, the effective number of replications, n_e, is:

$$n_e = \frac{16}{1 + f_A + f_F + f_C + f_E} = \frac{16}{5} = 3.2$$

Therefore, the confidence limit of $\hat{\mu}$ becomes:

$$\pm\sqrt{4.84 \times 3.73 \times \frac{1}{3.2}}$$

$$= \pm 2.38$$

It is expected to lie in the range between 9.87 and 14.63. The **confidence limit of the optimum condition,** when **the confirmatory trial is done with** r **experiments,** is given as follows:

$$\pm\sqrt{F_{fe}^{1}(0.05)\times V_e\left(\frac{1}{n_e}+\frac{1}{\mathrm{r}}\right)}$$

The previous calculation assumes that $r = \infty$.

The number of degrees of freedom for the factors of effective number of replications, n_e, used in the confidence limit for $\hat{\mu}$ is called the effective degrees of freedom. **The effective degrees of freedom of the dummy-level** is obtained as follows:

Effective degrees of freedom =

$$\frac{\text{Number of levels before taking the dummy of assigned factor}}{\text{Number of dummy} - \text{level duplication}} - 1$$

Table 8-7 lists the effective degrees of freedom of each level.

8-3. Formulae for Variation Calculation

It is necessary to calculate several variations in order to do analysis of variance. Variation calculations seem to be different for different assignment methods, e.g., the dummy-level technique, the combination technique, and the idle-column technique. Although the basic calculation method is simple, it requires a lot of experience to attain the level of application of a skilled engineer.

Furthermore, calculation of the effective degrees of freedom required to obtain the confidence limit for the optimum condition, $\hat{\mu}$, is rather complicated. *Table 8-6* and *Table 8-7* summarize the calculation formulas, so that even beginners can do calculations quickly and without difficulty.

8-4. ANOVA Examples for Various Assignment Techniques

This section gives the procedures for analysis of variance for the complex assignment techniques explained in Chapter 7. Variation calculation is done following *Table 8-6* and *Table 8-7*. The level sum and process average data used in analysis of variance is given in the supplementary tables in Chapter 7.

8-4-1. Example 1: Printed Circuit Board Test (L_{16}, Four Levels, Dummy-Level Technique, Idle-Column Technique Corresponds to subsection 7-8-1)

Variation Calculation and Analysis of Variance

1. Calculate factor variance and error variation

   Variation is calculated based on the level sums listed in *Table 7-12*:

$$ST = 44^2 + 33^2 + \cdots + 34^2 - \frac{559^2}{16} = 718.938$$

Table 8-6: Formulae for Total Variation (S_T), Between-Experiment Variation (S_{T_1}), and Repetition Variation (S_{e_2}):

Repetition	Variation Name	Variation Calculation	Degrees of Freedom	Variance
Yes ($r \geq 2$)	Total variation (S_T)	$S_T = y_{11}^2 + y_{12}^2 + \ldots y_{Lr}^2 - CF$	$f_T = N - 1$	—
	Between-exp. variation(S_{T_1})	$S_{T_1} = \frac{1}{r}\left(Y_1^2 + Y_2^2 + \ldots Y_L^2\right) - CF$	$f_{T_1} = L - 1$	—
	Repetition error variation (Second order error) (S_{e_2})	$S_{e_2} = S_T - S_{T_1}$	$f_{e_2} = f_T - f_{T_1}$ $= L \times (r - 1)$	Repetition variance (Second order error) $V_{e_2} = \frac{S_{e_2}}{f_{e_2}}$
No ($r = 1$)	Total variation (S_T)	$S_T = y_1^2 + y_2^2 + \ldots y_L^2 - CF$	$f_T = L - 1$ $= N - 1$	—

Table 8-7: Formulae for Factor Variation

Level	Orthogonal Array	Level Arrangement Method	Variation Name	Variation Calculation		Degrees of Freedom	Variance	Effective Degrees of Freedom	Note
2	2-Level system	Standard	Main effects	$S_A = \frac{1}{N}(A_1 - A_2)^2$ $= \frac{2}{N}\left(A_1^2 + A_2^2\right) - CF$		1	$V_A = \frac{S_A}{1}$	1	$(AB)_1 = A_1B_1$ $(AB)_2 = A_2B_1$ $(AB)_3 = A_2B_2$ $S_{AB} = \frac{3}{N}\left[(AB)_1^2 + (AB)_2^2 + (AB)_3^2\right] - CF$ $S_A = \frac{3}{2N}\left((AB)_1 - (AB)_2\right)^2$ $S_B = \frac{3}{2N}\left((AB)_2 - (AB)_3\right)^2$
	L_{18}	The 1st column							
	3-Level system	Dummy method	Main effects	$S_A = \frac{1}{2N}(A_1 + A_1' - 2A_2)^2$		1	$V_A = \frac{S_A}{1}$	$A_1 \Rightarrow 0.5$ $A_2 \Rightarrow 2.0$	
			Error	$S_{e_1} = \frac{1}{N} \times \frac{3}{2}(A_1 - A_1')^2$		1	$V_{e_1} = \frac{S_{e_1}}{1}$	—	
		Combining technique	Main effects	$S_A > S_B$	$S_A > S_B$				
				$S_A{}^* = \frac{3}{N}(AB)_1^2 + \frac{3}{2N}\left\{(AB)_2 + (AB)_3\right\}^2 - CF$	$S_B{}^{**} = \frac{3}{2N}\{(AB)_1 + (AB)_2\}^2 + \frac{3}{N}(AB)_3^2 - CF$	1	$V_A = \frac{S_A}{1}$ $V_B = \frac{S_B}{1}$	$A_2B_1 \Rightarrow 0.5$ $A_1B_2 \Rightarrow 2.0$	
				$S_B = \frac{3}{2N}\left\{(AB)_2 - (AB)_3\right\}^2$	$S_A = \frac{3}{2N}\left\{(AB)_1 - (AB)_2\right\}^2$	1	$V_B = \frac{S_B}{1}$ $V_A = \frac{S_A}{1}$	—	S_A and S_B may be calculated as: $^*S_A = S_{AB} - S_B$ $^{**}S_B = S_{AB} - S_A$

Table 8-7: Formulae for Factor Variation (Cont.)

Level	Orthogonal Array	Level Arrange-ment Method	Variation Name	Variation Calculation	Degrees of Freedom	Variance	Effective Degrees of Freedom	Note
3	3-level system	Standard	Main effects	$S_A = \frac{3}{N}\left(A_1^2 + A_2^2 + A_3^2\right) - CF$	2	$V_A = \frac{S_A}{2}$	2	
3	2-level system	Dummy method 4⇒3 levels	Main effects	$S_A = \frac{4}{N}A_1^2 + \frac{2}{N}(A_2 + A_2')^2 + \frac{4}{N}A_3^2 - CF$	2	$V_A = \frac{S_A}{2}$	$A_1, A_3 \Rightarrow 3$ $A_2 \Rightarrow 1$	
3	2-level system	Dummy method 4⇒3 levels	Error	$S_{e_1} = \frac{2}{N}(A_2 - A'_2)^2$	1	$V_{e_1} = \frac{S_{e_1}}{1}$	—	
3	2-level system	Idle column method	Main effects	$S_{A_1A_2} = \frac{2}{N}(A_1 - A_2)^2$	1	$V_{A_1A_2} = \frac{S_{A_1A_2}}{1}$	2	Obtain factor variation assigned to 1st and 2nd levels of the idle-column
3	2-level system	Idle column method	Main effects	$S_{A_2A_3} = \frac{2}{N}(A_2 - A_3)^2$	1	$V_{A_2A_3} = \frac{S_{A_2A_3}}{1}$	2	
4	2-level system	Multi-level	Main effects	$S_A = \frac{4}{N}\left(A_1^2 + A_2^2 + A_3^2 + A_4^2\right) - CF$	3	$V_A = \frac{S_A}{3}$	3	
6	L_{18}	"	"	$S_A = \frac{6}{N}\left(A_1^2 + A_2^2 + \ldots + A_5^2 + A_6^2\right) - CF$	5	$V_A = \frac{S_A}{5}$	5	
8	L_{16}, L_{32}	"	"	$S_A = \frac{8}{N}\left(A_1^2 + A_2^2 + \ldots + A_7^2 + A_8^2\right) - CF$	7	$V_A = \frac{S_A}{7}$	7	
9	L_{27}	"	"	$S_A = \frac{9}{N}\left(A_1^2 + A_2^2 + \ldots + A_8^2 + A_9^2\right) - CF$	8	$V_A = \frac{S_A}{8}$	8	

$$S_{(\text{idle})} = \frac{1}{8}\left(293^2 + 266^2\right) - (559)^2 / 16 = \frac{1}{16}(293 - 266)^2 = 45.563$$

$$\left.\begin{aligned} S_{A_1A_2} &= \frac{1}{8}(147 - 146)^2 = 0.125 \\ S_{A_2A_3} &= \frac{1}{8}(135 - 131)^2 = 2.00 \\ S_{B_1B_2} &= \frac{1}{8}(156 - 137)^2 = 45.125 \\ S_{B_2B_3} &= \frac{1}{8}(122 - 144)^2 = 60.500 \end{aligned}\right\} \text{Variations of } A \text{ and } B \text{ are calculated for different levels of the idle column}$$

$$S_C = \frac{1}{4}(137^2 + 140^2 + 143^2 + 139^2) - \frac{(559)^2}{16} = 4.688$$

$$S_D = \frac{1}{8}(131 + 141)^2 + \frac{1}{4}(146^2 + 141^2) - \frac{(559)^2}{16} = 17.188$$

$$S_E = \frac{1}{16}(284 - 275)^2 = 5.063$$

$$S_F = \frac{1}{16}(265 - 294)^2 = 52.563$$

$$S_G = \frac{1}{16}(323 - 236)^2 = 473.063$$

$$S_H = \frac{1}{16}(278 - 281)^2 = 0.563$$

$$S_{e_1} = \frac{1}{8}(131 - 141)^2 = 12.5 \text{ (from the dummy levels of D)}$$

S_D can also be calculated as follows:

$$S_D = \frac{1}{4}(146^2 + 131^2 + 141^2 + 141^2) - \frac{(559)^2}{16} - \frac{1}{8}(131 - 141)^2$$

$$= 17.188$$

The variation calculation results are summarized in ANOVA *Table 8-8.*

Table 8-8: ANOVA Table

Factor		f	S	V	F_o	Significance Test	Contribution Ratio (%)
$\bar{y}$		1	45.563	45.563	10.82	**	5.75
A	$A_1 A_2$	1	0.125	0.125			
	$A_2 A_3$	1	2.000	2.000			
B	$B_1 B_2$	1	45.125	45.125	10.71	**	5.69
	$B_2 B_3$	1	60.500	60.500	14.36	**	7.83
C		3	4.688	1.563			
D		2	17.188	8.594			
E		1	5.063	5.063			
F		1	52.563	52.563	12.48	**	6.73
G		1	472.063	472.063	112.30	**	65.21
H		1	0.563	0.563			
e_1		1	12.500	12.500			
T		15	718.938				100.00
(e)		(10)	(42.125)	(4.213)			(8.79)

Error variation is created by summing the variation that covers one-half to two-thirds of the total degrees of freedom.

$$S_e = S_{A_1A_2} + S_{A_2A_3} + S_C + S_D + S_E + S_H + S_{e_1} = 42.125$$

$$f_e = 1+1+3+2+1+1+1 = 10$$

$$V_e = \frac{S_e}{f_e} = 4.213$$

2. Calculate variance ratio

   The variance ratio, $F_{o,}$ is obtained using the following formula:

$$F_o = \left(\frac{\text{factor variance}}{\text{error variance}}\right)$$

3. Significance Test

   Those factors marked ** and * are significant at the 1% level and 5% level, respectively (*Table 8-8*).

4. Calculate contribution ratio

   The contribution ratio is calculated as follows:

$$\rho = \left[\frac{(S - f \times V_e)}{S_T}\right] \times 100\ (\%)$$

The contribution ratio of the idle column is:

$$\rho_{(\text{idle})} = \frac{(45.563 - 1 \times 4.213) \times 100}{718.938} = 5.75\ (\%)$$

Other contribution ratios are similarly calculated.

The confidence limit of the process averages is:

B_1, B_2, and B_3

$$\pm\sqrt{F_{10}^{1} \times \frac{V_e}{n_e}} = \pm\sqrt{4.96 \times \frac{4.213}{4}} = \pm 2.286$$

F and G

$$\pm\sqrt{F_{10}^{1} \times \frac{V_e}{n_e}} = \pm\sqrt{4.96 \times \frac{4.213}{8}} = \pm 1.616$$

Optimum Condition Estimate

Because the characteristic value is an LTB value, the levels of significant factors B_3, F_2, and G_1 are selected as the optimum condition. Its estimate, $\hat{\mu}$, is calculated as follows:

$$\begin{aligned}\hat{\mu}(B_3F_2G_1) &= \overline{T} + (\overline{B}_3 - \overline{T}) + (\overline{F}_2 - \overline{T}) + (\overline{G}_1 - \overline{T}) \\ &= \overline{B}_3 + \overline{F}_2 + \overline{G}_1 - 2 \times \overline{T} \\ &= 37.875 + 36.750 + 40.375 - 2 \times (\frac{559}{16}) \\ &= 45.215 \pm 2.555\end{aligned}$$

The effective number of replications, n_e, is:

$$n_e = \frac{16}{1 + 2\,(B) + 1\,(F) + 1\,(G)} = \frac{16}{5} = 3.2$$

The confidence limit is:

$$\pm\sqrt{F_{10}^{1} \times \frac{V_e}{n_e}} = \pm\sqrt{4.96 \times \frac{4.213}{3.2}} = \pm 2.555.$$

The optimum condition, judged from the factorial effects in Subsection 7-8-1, included factor level D_1. This gave a higher estimate for $\hat{\mu}$ as compared to the estimate based only on the significant factors. However, the optimum level combination is the same, because the best factor level must be chosen even if the factor is found to be nonsignificant.

8-4-2. Example 2: Surface Washing Test (L_9, Combining Technique, Dummy-Level Technique. Corresponds to Subsection 7-8-2)

Variation Calculation and Analysis of Variance

1. Calculate factor variance and error variance

   Variation is calculated based on *Table 7-15* and *Table 7-16*.

$$S_T = (-20.029)^2 + (-15.643)^2 + \cdots + (-15.441)^2 - \frac{(-173.807)^2}{9}$$

$$= 131.224 \qquad (f_T = 8)$$

$$S_A = \frac{1}{3}\{(-50.442)^2 + (-70.929)^2 + (-52.438)^2\} - \frac{(-173.807)^2}{9}$$

$$= 85.037 \qquad (f_A = 2)$$

$$S_{BC} = \frac{1}{3}\{(-65.993)^2 + (-57.576)^2 + (-50.241)^2\} - \frac{(-173.807)^2}{9}$$

$$= 41.435 \qquad (f_{BC} = 2)$$

$$S_D = \frac{1}{3}\{(-57.864)^2 + (-57.857)^2 + (-58.089)^2\} - \frac{(-173.807)^2}{9}$$

$$= 0.011 \qquad (f_D = 2)$$

$$S_E = \frac{1}{6}(-118.947)^2 + \frac{1}{3}(-54.863)^2 - \frac{(-173.807)^2}{9}$$

$$= 4.730 \qquad (f_E = 1)$$

$$S_{e_1} = \frac{1}{6}(E_1 - E_1')^2 = \frac{1}{6}\{(-59.597) - (-59.35)\}^2$$

$$= 0.010 \qquad (f_{e_1} = 1)$$

The best level combination is $(BC)_3 = B_2C_2$. Furthermore:

$$S_B = \frac{1}{6}[(BC)_2 - (BC)_3]^2 = \frac{1}{6}[(-57.576) - (50.241)]^2$$

$$= 8.967$$

$$S_C = \frac{1}{6}[(BC)_1 - (BC)_2]^2 = \frac{1}{6}[(-65.993) - (57.576)]^2$$

$$= 11.808$$

(BC) is treated as one factor in the analysis because the variation of B and C is about the same, and B_2 and C_2 are preferable levels according to the factorial effects. $(BC)_3 = B_2C_2$ is used as the third level in the experiment.

When S_{BC} is large and either S_B or S_C is small, the level of a factor with small variation is treated as the dummy-level. It is divided into large factor variation and error variation. For example, if S_C is large, the following recalculation is necessary before performing analysis of variance:

$$S_C = \frac{1}{3}(B_1C_1)^2 + \frac{1}{6}\{(B_1C_2)^2 + (B_2C_2)^2\} - \frac{T^2}{18}$$

$$S_B = \frac{1}{6}[(B_1C_2) - (B_2C_2)]^2$$

If S_{BC} is small, we can judge that no combination of BC will be effective. The sum of factor variation is 131.223, which is equal to the total variation, 131.224. The 0.001 difference is due to rounding. These results are summarized in ANOVA *Table 8-9*. The variance V_A of factor A is:

$$V_A = \frac{S_A}{f_A} = \frac{85.037}{2} = 42.519$$

The rest of the variances are calculated in the same way. A new error variance, V_e, is created by pooling D, E, and e_1, whose variances are small:

$$V_e = \frac{S_e}{f_e} = \frac{S_D + S_E + S_{e_1}}{f_D + f_E + f_{e_1}} = \frac{0.011 + 4.73 + 0.010}{2 + 1 + 1} = 1.188$$

2. Calculate variance ratio

   The variance ratio, F_o, of factor A is calculated as follows:

$$F_{o_A} = \frac{V_A}{V_e} = \frac{42.519}{1.188} = 35.79$$

3. Significance test

   As for the significance test, factors are marked with ** if their variance ratio, F_o, is larger than $F_4^2(0.01) = 18.00$, with * if it is between 18.00 and $F_4^2(0.05) = 6.94$, and with Δ if it is less than 6.94.

4. Calculate contribution ratio

   The contribution ratio for factor A is:

$$\rho_A = \frac{S_A - f_A \times V_e}{S_T} \times 100 = \frac{85.037 - 2 \times 1.188}{131.224} \times 100 = 62.99\ (\%)$$

The confidence limit of the process average of A and (BC) is:

$$\pm\sqrt{F_{fe}^1 \times \frac{V_e}{n_e}} = \pm\sqrt{7.71 \times \frac{1.188}{3}} = \pm 1.747.$$

Table 8-9: ANOVA Table

Factor	f	S	V	F_o	Signif. Test	Contribution Ratio
A	2	85.037	42.519	35.79	**	62.99
(BC)	2	41.435	20.717	17.44	*	29.76
D	2	0.011	0.006			
E	1	4.730	4.730			
e_1	1	0.010	0.010			
T	8	131.224				100.00
	(4)	(4.751)	(1.188)			(7.25)

Optimum Condition Estimation

Factors A and (BC) are significant and their optimum levels are A_1 and (B_2C_2), respectively. The estimate of the optimum condition, $\hat{\mu}$, is:

$$\hat{\mu}(A_1B_2C_2) = \overline{A}_1 + \overline{B_2C_2} - \overline{T} = -16.814 - 16.747 - (-19.312)$$

$$= -14.249 \quad [\text{db}]$$

The confidence limit is:

$$\pm\sqrt{F_4^1 \times \frac{V_e}{n_e}} = \pm\sqrt{7.71 \times \frac{1.188}{1.8}} = \pm 2.256.$$

The effective number of replications n_e, is:

$$n_e = \frac{9}{1+2+2} = \frac{9}{5} = 1.8$$

The range of existence is –16.505 ~ –11.993 [db] and this is 3.98 ~ 6.68 in terms of residual dirt quantity. The central value is estimated to be 5.16.

8-4-3. Example 3: Metal Surface Processing (L_{16}, Nested-Factor Technique. Corresponds with Subsection 7-8-3)

Variation Calculation and Analysis of Variance

Variation is calculated as follows:

$$S_T = 47^2 + 102^2 + \cdots + 56^2 - \frac{1043^2}{16} = 7486.4375$$

$$S_C = \frac{1}{8}(490^2 + 553^2) - \frac{1043^2}{16} = \frac{1}{16}(490 - 553)^2 = 248.0625$$

$$S_{D'} = \frac{1}{4}(279^2 + 211^2) - \frac{(279+211)^2}{8} = \frac{1}{8}(279-211)^2$$
$$= 578.0000$$

$$S_{D''} = \frac{1}{8}(300-253)^2 = 276.125$$

$$S_{E'} = 0.500, S_{E''} = 15.125$$

$$S_J = \frac{1}{16}(530-513)^2 = 18.0625$$

$$S_{K'} = 60.500, \quad S_{K''} = 325.125$$

$$S_{L''} = 50.000, \quad S_{L''} = 55.125$$

$$S_A = \frac{1}{8}(385^2 + 658^2) - \frac{1043^2}{16} = 4658.0625$$

$$S_B = 115.5625, \quad S_F = 689.0625$$

$$S_G = 351.5625, \quad S_H = 45.5625$$

The variances, variance ratios, significance test and contribution ratios are summarized in *Table 8-10*.

Table 8-10: ANOVA

Factor	f	S	V	F_0	Test	ρ (%)
C	1	248.0625	248.0625	7.09	*	2.85
D: D'	1	578.000	578.000	16.52	**	7.25
D''	1	276.125	276.125	7.89	*	3.22
E: E'	1	0.500	—	—		—
E''	1	15.125	—	—		—
J	1	18.0625	—	—		—
K: K'	1	60.500	—	—		—
K''	1	325.125	325.125	9.29	*	3.88
L: L'	1	50.000	—	—		—
L''	1	55.125	—	—		—
A	1	4658.0625	4658.0625	133.16	**	61.75
B	1	115.5625	115.5625	3.30	Δ	1.08
F	1	689.0625	689.0625	19.70	**	8.74
G	1	351.5625	351.5625	10.05	*	4.23
H	1	45.5625	—	—		—
T	15	7486.4375	7486.4375			100.00
(e)	7	(244.875)	(34.982)			(7.00)

The calculation for the confidence limit of the process average is:

$$C, A, B, F, G, \qquad \pm\sqrt{5.99 \times \frac{34.982}{8}} = \pm 5.12$$

$$D', D'', K'', \qquad \pm\sqrt{5.99 \times \frac{34.982}{4}} = \pm 7.24$$

Factorial effects are shown in *Figure 8-5*.

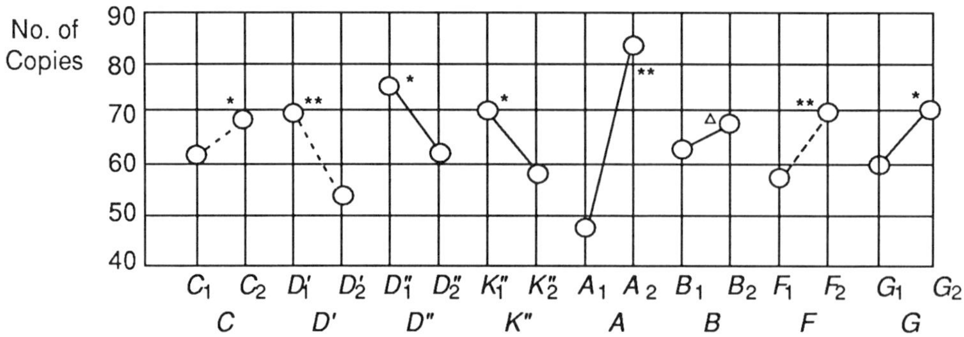

Figure 8-5: Factorial Effects

Optimum Condition Estimate

A key step in the nested-factor technique is to first determine the optimum conditions for both C (C_1, C_2) and J (J_1, J_2), compare them, and then select one.

The optimum condition of C_1 is C_1D_1' and that of C_2 is C_2D_1'':

$$\hat{\mu}(C_1D_1') = \overline{C}_1 + (\overline{D}_1' - \overline{C}_1) = \overline{D}_1' = \frac{279}{4} = 67.95 \pm 7.24$$

$$\hat{\mu}(C_2D_1'') = \overline{C}_2 + (\overline{D}_1'' - \overline{C}_2) = \overline{D}_1'' = \frac{300}{4} = 75.00 \pm 7.24$$

The confidence limit is calculated as follows:

$$\pm\sqrt{F_{fe}^1 \times \frac{V_e}{n_e}} = \pm\sqrt{5.99 \times 34.982 / 4} = \pm 7.24$$

$$\frac{1}{n_e} = \frac{1(m) + 1(C) + 2(D')}{\text{Total number of experiments}} = \frac{4}{16} = \frac{1}{4}$$

The effective degrees of freedom for D' is $(1 + 1) = 2$, because the estimation includes D and $C \times D$.

Under the optimum condition for C, the electrolyte method (C_2) is better. The significance difference between $\hat{\mu}(C_1D_1')$ and $\hat{\mu}(C_2D_1'')$ is determined as follows:

$$75.00 - 67.95 = 7.05$$

$$\sqrt{(7.24)^2 + (7.24)^2} = 10.24$$

Because the difference is smaller than the square root of the square sum of the existing limit, the difference is not significant.

The optimum condition for J is calculated as follows:

$$\hat{\mu}(J_1) = \frac{530}{8} = 66.25 \pm 5.12$$

$$\hat{\mu}(J_2K_1") = \frac{282}{4} = 70.50 \pm 7.24$$

The significance difference between $\hat{\mu}(J_1)$ and $\hat{\mu}(J_2K_1")$ is determined as follows:

$$70.50 - 66.25 = 4.25$$

$$\sqrt{(5.12)^2 + (7.24)^2} = 8.87$$

We can conclude that there is no significant difference.

The total optimum condition is determined from four combinations of C and J:

A. $\hat{\mu}(C_1D_1'J_1A_2B_2F_2G_2)$

$$= \overline{T} + (\overline{C_1D_1} - \overline{T}) + (\overline{J}_1 - \overline{T}) + (\overline{A}_2 - \overline{T}) + (\overline{B}_2 - \overline{T}) + (\overline{F}_2 - \overline{T}) + (\overline{G}_2 - \overline{T})$$

$$= \overline{C_1D_1} + \overline{J}_1 + \overline{A}_2 + \overline{B}_2 + \overline{F}_2 + \overline{G}_2 - 5\overline{T}$$

$$= 67.95 + 66.25 + \frac{658}{8} + \frac{548}{8} + \frac{574}{8} + \frac{559}{8} - 5 \times \left(\frac{1043}{16}\right)$$

$$= 100.125$$

B. $\hat{\mu}(C_1D_1''J_2K_1''A_2B_2F_2\,G_2)$

$$= 67.95 + 70.5 + 291.75 - 5 \times 65.1875$$

$$= 104.2625$$

C. $\hat{\mu}(C_2D_1''J_1A_2B_2F_2G_2)$

$$= 75.00 + 66.25 + 291.75 - 5 \times 65.1875$$

$$= 107.0625$$

D. $\hat{\mu}(C_2D_1''J_2K_1''A_2B_2F_2G_2)$

$$= 75.00 + 70.50 + 291.75 - 5 \times 65.1875$$

$$= 111.3125$$

The confidence limit of condition (D) is:

$$\text{Process confidence limit} = \pm\sqrt{F_{fe}^{1} \times \frac{V_e}{n_e}} = \pm\sqrt{5.99 \times 34.982 \times \left(\frac{10}{16}\right)}$$

$$= \pm 11.055$$

The effective number of replications is calculated as:

$$\frac{1}{n_e} = \frac{1(m)+1(C)+2(D'')+2(K'')+1(A)+1(B)+1(F)+1(G)}{16}$$

$$= \frac{10}{16}$$

8-5. Analysis of Variance for Classified Data (Accumulation Analysis) Introduction

Experimental data for engineering are divided into continuous data and noncontinuous or discrete data. Qualitative and ranked data are examples of discrete data. A typical evaluation method that yields qualitative data is the sensory test, in which the samples are compared with boundary samples or stepwise quality grade samples. The data are expressed as a discrete number of samples in each grade class. Even continuous values can be treated as discrete values if they are classified into different target levels. The **accumulation technique** is a method for analyzing data which are divided into more than three ranked classes.

8-5-1. Characteristic Value Classification and Accumulated Data

Characteristic values can be divided into six major categories:

1. Simple attributes
2. Simple variables
3. Classified attributes
4. Classified variables
5. Multi-attributes
6. Multi-variables.

Analysis based on the accumulation technique is applicable to (3) and (4), where data with ranking can be classified into more than three classes, e.g., good/average/acceptable and high/medium/low. Unranked classified data can be classified, but there is no order of significance, e.g., the types of cigarettes one likes or scissors/paper/stone.

The following examples of data can be analyzed with the accumulation technique.

1. Data from an appearance and sensory test
   - Appearance of a vinyl sheet
   - Taste and appearance of food

- Sensation of textile on skin
- Metal surface scratches.

2. Data containing continuous values and non-continuous values
   - Polymerization experiment data where some samples did not polymerize
   - Experiment data which have some unrotating motor samples
   - Experiment data which have overscaled data
   - Classified data based on stepwise quality grade samples and grade values.

Classification into four or five classes is recommended for the characteristic values listed above. The following symbols are generally used to indicate the classification:

◎: Most suitable to the purpose
O: Desirable
Δ: Acceptable
×: Not acceptable
××: Extremely far from the target

Treatment of data is required for analysis based on the accumulation technique. Suppose that data are given as ranked data classified into superior/good/acceptable/unacceptable, and that the number of samples in each class is three, four, one, and two, respectively. Cumulative frequencies must be determined. Data for Classes I, II, and III are analyzed, but data for Class IV are not.

Frequency

Superior	Good	Acceptable	Unacceptable
3	4	1	2

Cumulative Frequency

I	II	III	IV
Number of superior samples	Number of superior + good samples	Number of superior + good + acceptable samples	Number of superior + good + acceptable + unacceptable samples
3	7	8	10

Therefore, the number of classes that need to be analyzed is the original number of classes minus one.

8-5-2. Characteristics of the Accumulation Technique

The **accumulation technique** is a method for classifying data before analysis. **Its main feature is that it can evaluate the mean and the distribution at the same time.** The analysis method for continuous values described earlier allows us to evaluate the mean alone.

Suppose that three methods are used to produce a 5V constant-voltage battery without presupposing subsequent adjustment. Which method is the best? Intuitively, A_2 appears to be good, but all three methods have the same mean.

Production method	Data				Mean	Frequency			Cumulative frequency		
						O	Δ	×	I	II	III
A_1	4.2	5.8	5.1	4.9	5.0	2	0	2	2	2	4
A_2	5.0	5.1	4.9	5.0	5.0	4	0	0	4	4	4
A_3	4.0	6.2	4.5	5.3	5.0	0	1	3	0	1	4

Classification → O: 4.8 ~ 5.2, Δ: 4.5 ~ 4.7, ×: less than 4.4 or above 5.6

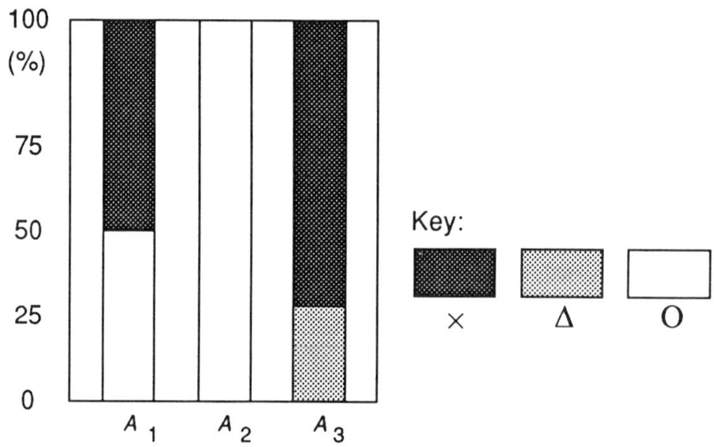

Figure 8-6: Factorial Effects

The factorial effects chart given in *Figure 8-6* is effective for selecting the levels. From the figure, it is clear that A_2 gives the target voltage, A_1 is the next best, and A_3 is the worst method. The traditional analytical method, which places emphasis on analysis of the mean, is incapable of selecting the best level based on information including dispersion. Level selection based on the accumulation technique enables selection of a level that has more samples in the preferable grade. Therefore, this method makes it possible to achieve the target value with small variability. This is true because the optimum condition is composed of levels that give samples concentrating around the target value. **Because the accumulation technique selects levels so that variability is controlled, this analysis method stabilizes the process and increases reliability.**

The accumulation technique examines the central value and dispersion simultaneously. The idea behind it is the same as that of the S/N ratio. The S/N ratio is defined as follows:

$$\text{S/N ratio} = \frac{\text{Signal}}{\text{Noise}}$$

The design concept that utilizes the S/N ratio is called parameter design. **The procedure for parameter design is as follows:**

Step 1. Select the combination of factor levels that gives the largest S/N ratio (the characteristic value output is usually different from the target value. The combination that minimizes variability must be chosen).

Step 2. Adjust the difference between the output and target value using a factor level that changes the central value considerably without influencing the S/N ratio.

An important task of design is to choose parameters that achieve the output at the target value with small variability. Because the difference between the output and the target value results in a loss, the accumulation technique and parameter design are important analysis techniques for minimizing loss. In the factorial effects chart, the pattern indicates the existence of the classified value. The length of each pattern indicates the dispersion, with the longer pattern meaning a concentrated distribution.

The power of the test will be greater if continuous data rather than classified data are used. For example, data for on/off switching is [0, 1] data, whose test power is low. In such a case, it is better to perform parameter design by using other output characteristics relating to the on/off function that can be treated as continuous data, such as temperature, displacement, and resistance. These characteristic values are called **secondary** or **alternative characteristics**. Continuous data are better for the design stage. Engineers should always try to quantify data.

Next, Subsection 8-5-3 introduces examples of the application of the accumulation technique. The entire analysis procedure, including variation calculation, analysis of variance, the drawing of the factorial effects chart, and optimum condition estimation is done using a readily available computer program.

8-5-3. An Example of the Accumulation Technique (L_{16}, Sensor Design)

In this example, the accumulation technique was successfully applied in new product development.

Background of the Experiment

This example relates to an office automation machine's electrical signal processing. The pulse signal sent by the transmitter needs to be received with only a small time lag, so as to speed up signal processing. The time lag is a Smaller-the-Better (STB) characteristic, because it should be as small as possible.

Factor Listing and Assignment

Several engineers listed about 30 factors. Nine factors, which seemed to be related to the signal processing speed and should have an improvement effect, were selected for the experiment. Wide levels were intentionally taken and factors were assigned to an L_{16}.

The following is a summary of the tested factors.

Three level factors:

Sensor main body	*C, A*
Sensor position	*K, F*

Two level factors:

Sensor shape	*B, E*
	G, I
	J

Figure 8-7 shows the a linear graph and *Table 8-11* shows the experimental results.

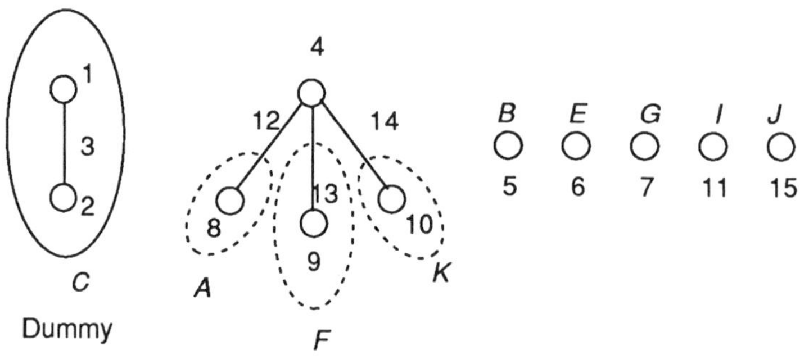

Figure 8-7: Linear Graph

Descriptions of the actual levels are omitted. Three levels were arranged by the dummy-level technique for factor *C*, and by the idle-column technique for factors *A*, *K*, and *F*. First order error variation, e_1, was calculated from the replicated dummy level.

Handling the Experimental Results

Among the 16 samples from the experiment, 11 could be measured as quantitative data. However, the signal was not detected for the remaining five samples. In other words, the signal was sent, but it was not detected.

There are many examples where qualitative data, such as "unmeasurable" data, appear among continuous data, as in this example. For instance about half of the samples of an LSI circuit formation experiment, which took place in the United States, failed in etching; 11 motors out of 32 from an L_{32} (one replication) did not rotate; one-third of the samples in an emulsion processing experiment using an L_{16} did not form an emulsion. In such cases, because conditions that should not appear in the product are created at the stage of experimentation, engineers may consider the experiment to be a failure and

Table 8-11: Assignment and Experimental Results

No.	C 1,2,3	Idle	A 8, 12	F 9, 13	K 10, 14	B 5	E 6	G 7	I 11	J 15	Raw Data	Classification	Frequency ◎	○	Δ	X	Cumulative Frequency I	II	III	IV
1	1	1	1	1	1	1	1	1	1	1	—	X	0	0	0	1	0	0	0	1
2	1	1	2	2	2	1	1	1	2	2	7.0	Δ	0	0	1	0	0	0	1	1
3	1	2	2	2	2	2	2	2	1	2	—	X	0	0	0	1	0	0	0	1
4	1	2	3	3	3	2	2	2	2	1	—	X	0	0	0	1	0	0	0	1
5	2	1	1	1	2	1	2	2	2	2	2.5	◎	1	0	0	0	1	1	1	1
6	2	1	2	2	1	1	2	2	1	1	4.0	○	0	1	0	0	0	1	1	1
7	2	2	2	2	3	2	1	1	2	1	4.2	○	0	1	0	0	0	1	1	1
8	2	2	3	3	2	2	1	1	1	2	7.0	Δ	0	0	1	0	0	0	1	1
9	3	1	1	2	1	2	1	2	2	2	7.0	Δ	0	0	1	0	0	0	1	1
10	3	1	2	1	2	2	1	2	1	1	6.3	Δ	0	0	1	0	0	0	1	1
11	3	2	2	3	2	1	2	1	2	1	—	X	0	0	0	1	0	0	0	1
12	3	2	3	2	3	1	2	1	1	2	2.5	◎	1	0	0	0	1	1	1	1
13	2'	1	1	2	2	2	2	1	1	1	—	X	0	0	0	1	0	0	0	1
14	2'	1	2	1	1	2	2	1	2	2	1.0	◎	1	0	0	0	1	1	1	1
15	2'	2	2	3	3	1	1	2	1	2	1.2	◎	1	0	0	0	1	1	1	1
16	2'	2	3	2	2	1	1	2	2	1	5.8	Δ	0	1	0	0	0	1	1	1

think "Why couldn't I think of factors and levels that would not give inferior samples in the experiment?" However, this is the wrong way to think about the problem.

Both good and inferior samples in experimental results are proof of a successful experiment. Good and inferior samples make it possible to select levels that produce only good products. Therefore, factor levels must be set widely so that both good and inferior samples are produced, and main effect factors must be assigned to all columns so that many improvement possibilities are tested.

Now, how should we handle the results of this example? There are two typical approaches.

- Substitute appropriate numbers for the five unmeasured data and analyze them as continuous data.
- Classify 16 data points into different rankings and analyze them as classified data based on the accumulation technique.

Although continuous data analysis has a high test power, substitution of arbitrary numbers to the cases of the experiment that failed to receive the signal is a distortion of reality. Level selection may become arbitrary and the optimum condition and its estimate will be unreliable. Experimental results should be classified for analysis according to the accumulation technique.

Limit	Grade	Classification Symbol
0.0–3.0	Best	◎
3.1–6.0	Good	O
6.1–10.0	Acceptable	Δ
10.1–	Unacceptable	X

The results are classified as in the preceding table and the frequency and cumulative frequency are obtained for each number. The results are shown in *Table 8-11*. Next, the level sum and average are calculated as shown in *Table 8-12*. The process average was obtained by dividing the level sum by the total number of data in each class and multiplying by 100.

Variation Calculation

After obtaining cumulative frequencies, the variation of Classes I, II, and III was calculated. The total variation was calculated for the overall evaluation by multiplying the **weight, *W*,** by the variation of each class. Calculations for Class IV are not necessary.

Weights W_I, W_{II}, and W_{III} are given as the inverse of the variance from the binomial distribution, where p_I, p_{II}, and p_{III} are the percentages of items falling into Classes I, II, and III, respectively.

Table 8-12: Level Sum and Average

	Level	Level Sum				Average (%)		
Factor	*Class*	1	2	3	Total	1	2	3
C	I	0	3	1	4	0.0	37.5	25.0
C	II	0	6	1	7	0.0	75.0	25.0
C	III	1	7	3	11	25.0	87.5	75.0
Idle	I	2	2		4	25.0	25.0	
Idle	II	3	4		7	37.0	50.0	
Idle	III	6	5		11	75.5	62.5	
A A_1A_2	I	1	1		2	25.0	25.0	
A A_1A_2	II	1	2		3	25.0	50.0	
A A_1A_2	III	2	4		6	50.0	100.0	
A A_2A_3	I		1	1	2		25.0	25.0
A A_2A_3	II		2	2	4		50.0	50.0
A A_2A_3	III		2	3	5		50.0	75.0
F F_1F_2	I	2	0		2	50.0	0.0	
F F_1F_2	II	2	1		3	50.0	25.0	
F F_1F_2	III	3	3		6	75.0	75.0	
F F_2F_3	I		1	1	2		25.0	25.0
F F_2F_3	II		3	1	4		75.0	25.0
F F_2F_3	III		3	2	5		75.0	50.0
K K_1K_2	I	1	1		2	25.0	25.0	
K K_1K_2	II	2	1		3	50.0	25.0	
K K_1K_2	III	3	3		6	75.0	75.0	
K K_2K_3	I		0	2	2		0.0	50.0
K K_2K_3	II		1	3	4		25.0	75.0
K K_2K_3	III		2	3	5		50.0	75.0
B	I	3	1		4	37.5	12.5	
B	II	5	2		7	62.5	25.0	
B	III	6	5		11	75.0	62.5	

Table 8-12 (Cont'd): Level Sum and Average

Factor	Level Class	Level Sum 1	2	3	Total	Average (%) 1	2	3
E	I	1	3		4	12.5	37.5	
E	II	3	4		7	37.5	50.0	
E	III	7	4		11	87.5	50.0	
G	I	2	2		4	25.0	25.0	
G	II	3	4		7	37.5	50.0	
G	III	5	6		11	62.5	75.0	
I	I	2	2		4	25.0	25.0	
I	II	3	4		7	37.5	50.0	
I	III	5	6		11	62.5	75.0	
J	I	0	4		4	0.0	50.0	
J	II	3	4		7	37.5	50.0	
J	III	4	7		11	50.0	87.5	

$$W_{\text{I}} = \frac{1}{p_{\text{I}}(1-p_{\text{I}})} = \frac{1}{\frac{4}{16}\left(1-\frac{4}{16}\right)} = \frac{16^2}{4(16-4)}$$

$$= 5.333$$

$$W_{\text{II}} = \frac{1}{p_{\text{II}}(1-p_{\text{II}})} = \frac{1}{\frac{4}{16}\left(1-\frac{7}{16}\right)} = \frac{16^2}{7(16-7)}$$

$$= 4.063$$

$$W_{\text{III}} = \frac{1}{p_{\text{III}}(1-p_{\text{III}})} = \frac{1}{\frac{11}{16}\left(1-\frac{11}{16}\right)} = \frac{16^2}{11(16-11)}$$

$$= 4.655$$

In this case, the number of **degrees of freedom for the accumulation technique** is the product of the original number of degrees of freedom and the number of classes to be analyzed, because the degrees of freedom correspond to Classes I, II, and III.

The total variation is as follows:

$$S_T = \left(\frac{0^2 + 0^2 + 0^2 + \cdots + 1^2 + 1^2 + 0^2 - 4^2}{16} \right) \times W_{\mathrm{I}}$$

$$+ \left(\frac{0^2 + 0^2 + 0^2 + \cdots + 1^2 + 1^2 + 1^2 - 7^2}{16} \right) \times W_{\mathrm{II}}$$

$$+ \left(\frac{0^2 + 1^2 + 0^2 + \cdots + 1^2 + 1^2 + 1^2 - 11^2}{16} \right) \times W_{\mathrm{III}}$$

$$= \left(4 - \frac{4^2}{16} \right) \times \frac{16^2}{4(16-4)} + \left(7 - \frac{7^2}{16} \right) \times \frac{16^2}{7(16-7)}$$

$$+ \left(11 - \frac{11^2}{16} \right) \times \frac{16^2}{11(16-11)}$$

$$= 16 + 16 + 16 = 16 \times 3 = 48$$

Total variation, S_T, is always a simple number.
The **total variation** is:

S_T = Total number of data × number of analyzed classes

The **variation of each factor** is calculated as follows:

$$S_C = \frac{1}{4} \left(0^2 + 3^2 + 0^2 - \frac{4^2}{16} \right) \times W_{\mathrm{I}}$$

$$+ \frac{1}{8} \left(0^2 + 6^2 + 1^2 - \frac{7^2}{16} \right) \times W_{\mathrm{II}}$$

$$+ \frac{1}{4} \left(1^2 + 7^2 + 3^2 - \frac{11^2}{16} \right) \times W_{\mathrm{III}}$$

$$= 13.083$$

First order error variation, S_{e_1}, is calculated from the dummy level:

$$S_{e_1} = \frac{1}{8}(1-2)^2 \times W_{\mathrm{I}} + \frac{1}{8}(3-3)^2 \times W_{\mathrm{II}} + \frac{1}{8}(4-3)^2 \times W_{\mathrm{III}}$$

$$= 1.248$$

$S_{A_1A_2}$ is calculated:

$$S_{A_1A_2} = \frac{1}{8}(1-1)^2 \times W_I + \frac{1}{8}(1-2)^2 \times W_{II} + \frac{1}{8}(2-4)^2 \times W_{III}$$

$$= 2.835$$

The rest are similarly calculated.

Analysis of Variance

1. Calculate variance and error variance

The number of degrees of freedom for each factor is obtained by multiplying by three, which is the number of analyzed classes. The variance is obtained by dividing the variation by the degrees of freedom:

$$V_C = \frac{13.803}{6} = 2.300$$

$$V_{(\text{Idle})} = \frac{0.545}{3} = 0.182$$

The rest are similarly calculated.

Here, the new error variance, V_e, is obtained from factors whose variance is less than 0.800:

$$\text{Error variation:} \quad S_e = S_{(\text{idle})} + S_{A_2A_3} + S_{K_1K_2} + S_G + S_I + S_{e_1}$$

$$= 3.973$$

$$\text{Errors degrees of freedom:} \quad f_e = 3+3+3+3+3+3 = 18$$

$$\text{Error variance:} \quad V_e = \frac{S_e}{f_e}$$

$$= \frac{3.973}{18} = 0.221$$

2. Calculate variance ratio

The variance ratio, F_o, is obtained by dividing unpooled factor variance by the new error variance:

$$\text{Variance ratio of factor } C = \frac{2.300}{0.221} = 10.42$$

$$\text{Variance ratio of } A_1A_2 = \frac{0.945}{0.221} = 4.28$$

The rest are similarly obtained.

3. Significance test

The significance test is done with the variance ratio and the degrees of freedom of each factor and that of the error term, which is 18. Factors are marked with ** and * if they are significant at the 1% and 5% significance level, respectively.

4. Calculate contribution ratio

The contribution ratio is determined. The following is the contribution ratio for factor C in the example:

$$\rho_C(\%) = \frac{(13.803 - 6 \times 0.221) \times 100}{48.00}$$

$$= 26.00(\%)$$

The contribution ratio of the error term is obtained by subtracting the sum of the contribution ratios of significant factors from 100:

$$\rho_e = 100 - (26.00 + 4.53 + \cdots + 15.72) = 20.69(\%)$$

The results of these calculations are summarized in ANOVA *Table 8-13*. The factorial effects chart is shown in *Figure 8-8*.

Table 8-13: ANOVA Table

Factor		f	S	V	F_0	Test	ρ (%)
C		6	13.803	2.300	10.42	**	26.00
Idle		3	0.545	0.182			
A	A_1A_2	3	2.835	0.945	4.28	*	4.53
	A_2A_3	3	0.582	0.194			
F	F_1F_2	3	3.175	1.058	4.79	*	5.23
	F_2F_3	3	2.614	0.871	3.95	*	4.07
K	K_1K_2	3	0.508	0.169			
	K_2K_3	3	5.280	1.760	7.97	**	9.62
B		3	3.910	1.303	5.90	**	6.77
E		3	4.205	1.402	6.35	**	7.38
G		3	0.545	0.182			
I		3	0.545	0.182			
J		3	8.205	2.735	12.39	**	15.72
e_1		3	1.248	0.416			
e_2		0	0.000	0.000			
T		45	48.000				100.00
(e)		(18)	(3.973)	(0.221)			(20.69)

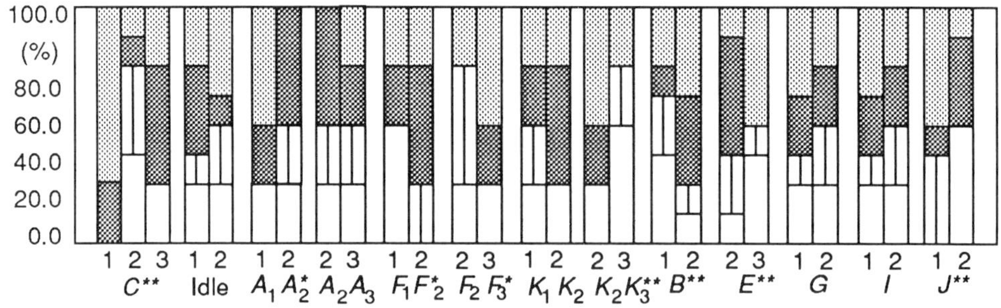

Figure 8-8: Factor Effect

Range of Existence and Estimate of the Optimum Condition

According to the factorial effects chart, the optimum condition is C_2, A_2, F_1, K_3, B_1, E_2, and J_2. The estimate of optimum combination is calculated for each class. $\hat{\mu}$ is estimated for each class using the **omega method**:

$$\hat{\mu}_{\mathrm{I}} = \overline{C}_2[\mathrm{db}] + \overline{A}_2[\mathrm{db}] + \overline{F}_1[\mathrm{db}] + \overline{K}_3[\mathrm{db}] + \overline{B}_1[\mathrm{db}] + \overline{E}_2[\mathrm{db}] + \overline{J}_2[\mathrm{db}] - 6 \times T[\mathrm{db}]$$

$$= 17.201\,[\mathrm{db}] = 98.130(\%)$$

$$\hat{\mu}_{\mathrm{II}} = 18.310\,[\mathrm{db}] = 98.546(\%)$$

$$\hat{\mu}_{\mathrm{III}} = 22.431\,[\mathrm{db}] = 99.432(\%)$$

Detailed calculation results are summarized in *Table 8-14*.

The **confidence limit** is obtained by considering the error variance, σ^2, of appearance frequency, p, to be proportional to $p\,(1-p)$.

$$\mu = \hat{\mu} \pm \sqrt{F_{fe}^{1}(0.05) \times V_e \hat{\mu}(1-\hat{\mu}) / n_e}$$

where n_e is the **effective number of replications**.

$$n_e = \frac{\text{total number of data}}{1 + \text{effective degrees of freedom of significant level}}$$

$$= \frac{16}{1+1+2+2+2+1+1+1} = \frac{16}{11}$$

Because the above equation gives the upper limit of the confidence limit, which is above 100%, the confidence limit should be obtained from the minute equation:

$$\frac{4.343}{\overline{T}(1-\overline{T})}\sqrt{F_{fe}^{1} \times V_e \times \frac{\overline{T}(1-\overline{T})}{n_e}}\ [\mathrm{db}]$$

Table 8-14: Optimum Combination Estimates

Factor	Level	Equation	Calculation Unit I	II	III
C	2	$1 * C_2$	–2.218	4.771	8.451
Idle					
A: A_1A_2	2	$1 * A_1$	–4.771	0.000	11.761
A: A_2A_3					
F: F_1F_2	1	$1 * F_1$	0.000	0.000	4.771
F: F_2F_3					
K: K_1K_2					
K: K_2K_3	3	$1 * K_2$	0.000	4.771	4.771
B	1	$1 * B_1$	–2.218	2.218	4.771
E	2	$1 * E_2$	–2.218	0.000	0.000
G					
I					
J	2	$1 * J_2$	0.000	0.000	8.451
Nested		Modified portion	0.000	0.000	0.000
T bar		–6 * T	28.627	6.549	–20.545
Combination estimate (db)			17.201	18.310	22.431
Combination estimate (%)			98.130	98.546	99.432

The confidence limit of Class I is the following:

$$\pm\frac{4.343}{\frac{4}{16}\left(1-\frac{4}{16}\right)}\times\sqrt{441\times 0.221\times\frac{\frac{4}{16}\left(1-\frac{4}{16}\right)}{\left(\frac{16}{11}\right)}}$$

$$=\pm\ 8.209\ [\text{db}]$$

$$\mu_{\text{I}} = 17.20 \pm 8.209$$

$$= 8.992 \sim 25.41\ [\text{db}]$$

$$= 88.800 \sim 99.713\ (\%)$$

μ_{II} and μ_{III} are similarly calculated and shown in *Table 8-15*.

Table 8-15: Confidence Limit of Optimum Condition

Class	Combination	+/– (db)	Range (%)
I	98.130	8.209	88.800.....99.713
II	98.546	7.165	92.866.....99.717
III	99.432	7.669	96.768.....99.902

Confirmatory Trial

Mass production with the optimum condition showed the level of time lag to be less than 0.8. Thus, the initial quality target was achieved.

8-5-4. Conclusion

This subsection has discussed the accumulation technique, which is used to analyze experimental results containing qualitative data. The technique classifies the samples into more than three rankings. This method gives information that improves both the mean and the variability simultaneously. It is effective for improving the stability and reliability required in both instruments and processes.

The so-called inferior products produced at the research level do not represent a loss. About half of the samples in an experiment of new product development should be nonfunctional products. A substantial loss will result, however, if inferior products are produced in actual production. In order to avoid such a loss, it is necessary to take many factors together with wide factor levels, which may create nonfunctional products. Nonfunctional products are treated as qualitative data and classified for analysis using the accumulation technique.

There are several data analysis methods. These include the accumulation technique and the S/N ratio method. However, the most important thing for engineers to understand is the importance of their experiments and how to make appropriate judgments about which information, of what quality, is needed.

8-6. Analysis of Variance for Classified Data, ([0, 1] Data Corresponds with the Example in Table 2-9)

The characteristic values that are classified into two classes are called **[0, 1] data**. Examples of two-class classifications are good/inferior, OK/NG, GO/NO, up/down, and right/left. In this section, the good/inferior data from appearance judgment, given in *Table 2-9*, are analyzed as [0, 1] data.

8-6-1. Variation Calculation

Numbers 1 and 0 are assigned to good and inferior samples determined by appearance judgment. Level sums A_1, A_2, G_1, and G_2 are calculated as follows:

$$A_1 = 4+5+5+2 = 16$$

$$A_2 = 2+0+2+0 = 4$$

$$G_1 = 4+2+0+2 = 8$$

$$G_2 = 5+5+2+0 = 12$$

Other level sums are similarly calculated and are summarized in *Table 8-16.*

Table 8-16: Level Sum (Average)

Factor	1	2
A	16(80)	4(20)
B	11(55)	9(45)
C	11(55)	9(45)
D	13(65)	7(35)
E	9(45)	11(55)
F	8(40)	12(60)
G	8(40)	12(60)

Numbers in parentheses are percentages (%)

The total number of experiments is 40 and each **variation** is determined as follows:

$$S_m = \text{CF} = \frac{(\text{Total number of samples classified as 1})^2}{\text{Number of data}} = \frac{20^2}{40} = 10$$

$$(f_m = 1)$$

$$S_T = \text{Square sum of each datum} - \text{CF}$$

$$= 20 - 10$$

$$= 10 \quad (f_T = 40 - 1 = 39)$$

$$S_{T_1} = \frac{1}{5}(4^2 + 5^2 + 5^2 + \cdots + 0^2) - \text{CF}$$

$$= 5.6 \quad (f_{T_1} = 8 - 1 = 7)$$

$$S_{e_2} = S_T - S_{T_1} = 10 - 5.6 = 4.4 \quad (f_{e_2} = f_T - f_{T_1} = 32)$$

$$S_A = \frac{(16-4)^2}{40} = 3.6 \quad (f = 1)$$

$$S_B = \frac{(11-9)^2}{40} = 0.1 \quad (f = 1)$$

$$S_C = \frac{(11-9)^2}{40} = 0.1 \quad (f = 1)$$

$$S_D = \frac{(13-7)^2}{40} = 0.9 \quad (f = 1)$$

$$S_E = \frac{(9-11)^2}{40} = 0.1 \quad (f = 1)$$

$$S_F = \frac{(8-12)^2}{40} = 0.4 \quad (f = 1)$$

$$S_G = \frac{(8-12)^2}{40} = 0.4 \quad (f = 1)$$

Because the sum of factor variation equals S_{T_1}, the calculation is correct.

$$\begin{aligned} \text{Column variation sum} &= S_A + S_B + \cdots + S_G \\ &= 3.6 + 0.1 + \cdots + 0.4 \\ &= 5.6 \end{aligned}$$

8-6-2. Analysis of Variance

Variance is obtained by dividing the variation by the degrees of freedom.

The **second order error variance,** V_{e_2} becomes the following:

$$V_{e_2} = \frac{S_{e_2}}{f_{e_2}} = \frac{4.4}{3.2} = 0.1375$$

These results are summarized in ANOVA *Table 8-17*. A **new error variance,** V_e, is obtained by pooling the factor variation of factors having small variance:

$$V_e = \frac{S_B + S_C + S_E + S_F + S_G + S_{e_2}}{f_B + f_C + f_E + f_F + f_G + f_{e_2}} = \frac{5.5}{37}$$

$$= 0.1486$$

Although it is difficult to decide whether factors F and G should be pooled, they are pooled in this analysis of variance. However, no loss will be generated by this procedure, because the better level must be selected even for the non-significant factors.

The **variance ratio,** F_o, will be:

$$F_{0m} = \frac{V_m}{V_e} = \frac{10.00}{0.1486} = 67.29$$

$$F_{0A} = \frac{3.6}{0.1486} = 24.23$$

$$F_{0D} = \frac{0.9}{0.1486} = 6.06$$

As for the **significance test,** because F_{37}^{1} is not shown in the F Table, we use F_{36}^{1}, whose 1% value is 7.39 and 5% value is 4.11. We mark those factors with ** whose F_o is greater than 7.39 and with * if F_o lies between 7.39 and 4.11.

The **contribution ratio** is calculated as follows:

$$\rho_m = \frac{S - f \times V_e}{S_T} \times 100 = \frac{10 - 1 \times 0.1486}{20} \times 100$$

$$= 49.26 \quad (\%)$$

The rest of the results are similarly derived. They are summarized in *Table 8-17.*

Table 8-17: ANOVA Table

Factor	f	S	V	F_o	Test	ρ (%)
	1	10.0	10.0	67.29	**	49.26
A	1	3.6	3.6	24.23	**	17.26
B	1	0.1	—	—		—
C	1	0.1	—	—		—
D	1	0.9	0.9	6.06	*	3.76
E	1	0.1	—			—
F	1	0.4	—			—
G	1	0.4	—			—
e_2	32	4.4	0.1375			—
T	40	20.0				100.00
(e)	(37)	(5.5)	(0.1486)			(29.72)

8-6-3. Optimum Condition and Confidence Limit

The **confidence limit of the process** in the case of $\overline{A}_2 = 20\%$ is the following:

$$\pm\sqrt{F_{fe}^{1} \times \frac{\overline{A}_2(1-\overline{A}_2)}{\overline{T}(1-\overline{T})} \times \frac{V_e}{n_e}} = \sqrt{4.11 \times \frac{0.2(1-0.2)}{0.5(1-0.5)} \times \frac{0.1486}{20}}$$

$$= \pm 14.0\,(\%)$$

where $\overline{A}_2 = 0.2$, $\overline{T} = 0.5$, $n_e = 20$, and $V_e = 0.1486$.

The rest are similarly calculated.

The optimum condition is A_1 and D_1:

$$\hat{\mu}(A_1D_1) = \overline{A}_1(80\%) + \overline{D}_1(65\%) - \overline{T}(50\%)$$

$$= 6.021 + 2.688 - 0$$

$$= 8.709\ [\text{db}]$$

$$= 88.1\,(\%)$$

The **confidence limit of the optimum condition** is derived using the minute equation explained in the section of on accumulation analysis:

$$\frac{0.343}{0.5(1-0.5)} \times \sqrt{4.11 \times 0.1486 \times 0.5(1-0.5) \times \frac{3}{40}}$$

$$= 1.8590 \quad [\text{db}]$$

The **range of existence** becomes:

$$8.709 \pm 1.8590 = 6.850 \sim 10.568 \quad [\text{db}]$$

$$= 82.9 \sim 91.9 \quad (\%)$$

With [0, 1] data, S_m must also be analyzed. This is because a target of "good" is 100%. The target is zero for residual grease and defective ratio. Therefore, it is also necessary to know how far away the central value of the experiment lies from the target. Because the contribution ratio of average m is 49.26 in this example, it is expected that achieving the 100% target will be difficult.

8-7. Examples of Analysis of Variance

This sub-section discusses the analysis of variance of experimental design examples according to the manual. The description is limited to particular points of interest, including the derivation of variation.

8-7-1. Analysis of Variance for Variability Reduction, Research on Optical-Part Hole Position (Corresponds to 6-1)

Analysis of Variance for Hole Position

Variation is calculated as follows:

$$\text{Total variation} \quad S_T = (-24.937)^2 + (-21.548)^2 + (-22.439)^2 + \cdots + (-22.438)^2 - \frac{T^2}{12} = 108.123$$

$$(T = -24.937 - 21.548 - \cdots - 22.438 = -296.78)$$

The number of degrees of freedom is the number of data minus one, i.e., $f_T = 12 - 1 = 11$.

The **variation of factor** A, S_A, is:

$$S_A = \frac{1}{12}\{-154.21 - (-142.57)\}^2 = 11.295$$

and its degrees of freedom, f_A, is the number of levels minus one, i.e., $f_A = 2 - 1 = 1$. The variation and degrees of freedom for the other factors are similarly calculated.

Because the computer output of level sums is limited to two digits below zero, the number of the third digit below zero for variation may be different. The results are shown in *Table 8-18*. **Variance** is obtained by dividing variation by the degrees of freedom:

$$V_A = \frac{S_A}{f_A} = 11.295$$

$$V_B = \frac{S_B}{f_B} = 4.004$$

The variance for the other factors is similarly calculated.

New error variance, V_e, is calculated by pooling variations of small variance factors, i.e., S_C, S_F, S_G, S_H, and S_I:

$$V_e = \frac{S_C + S_F + S_G + S_H + S_I}{f_C + f_F + f_G + f_H + f_I} = \frac{1.399}{5} = 0.280$$

The **variance ratio** is obtained by dividing unpooled factor variance, V, by new error variance, V_e.

$$F_{0A} = \frac{V_A}{V_e} = \frac{11.295}{0.280} = 40.38$$

As for the **significance test**, the factors are marked with ** if their variance ratio is greater than $F_5^1(0.01) = 16.26$, with * if between 16.26 and $F_5^1(0.05) = 0.61$, and with D if less than 6.61.

The **contribution ratio** of factor A, ρ_A, is calculated as follows.

$$\rho_A = \frac{S_A - f_A \times V_e}{S_T} \times 100 = \frac{11.295 - 1 \times 0.280}{108.123} \times 100 = 10.19\ (\%)$$

The contribution ratios for the rest of the factors are similarly calculated.

The **contribution ratio of the error term** is small:

$$\rho_e = 100 - (\rho_A + \rho_B + \rho_D + \rho_E + \rho_J + \rho_K) = 2.85\ (\%)$$

This indicates that the noise component of the experiment was 2.85%, and 97.15% is caused by the factorial effects of factors assigned in the experiment. This means that the system is easy to control, and that it is possible to create an ideal system.

The **confidence limit of the process average** is:

$$\pm\sqrt{F_5^1 \times \frac{V_e}{n_e}} = \pm\sqrt{6.61 \times \frac{0.280}{6}} = \pm 0.555$$

The **effective number of replications**, n_e, is given as:

$$n_e = \frac{12}{1 + f_A + f_B + f_D + f_E + f_J + f_K} = \frac{12}{7}$$

and the **confidence limit for the optimum condition** is:

$$\pm\sqrt{F_5^1 \times \frac{V_e}{n_e}} = \pm\sqrt{6.61 \times \frac{0.280}{\left(\frac{12}{7}\right)}} = \pm 1.04$$

Table 8-18: ANOVA Table—Hole Position

Factor	f	S	V	F_0	Test	ρ(%)
A	1	11.295	11.295	40.38	**	10.19
B	1	4.004	4.004	14.31	*	3.44
C	1	0.441	0.441			
D	1	43.495	43.495	155.49	**	39.97
E	1	21.046	21.046	75.24	**	19.21
F	1	0.637	0.637			
G	1	0.009	0.009			
H	1	0.006	0.006			
I	1	0.307	0.307			
J	1	17.661	17.661	63.14	**	16.08
K	1	9.223	9.223	32.97	**	8.27
T	11	108.123				100.00
(e)	(5)	(1.399)	(0.280)			(2.85)

Analysis of Variance for Burr Generation

Weights W_I and W_{II} of Class I and II are calculated as follows:

$$W_I = \frac{1}{p_I(1-p_I)} = \frac{1}{\frac{p_I}{T}(1-\frac{p_I}{T})} = \frac{T^2}{p_I(T-p_I)}$$

$$= \frac{240^2}{130(240-130)} = 4.028$$

$$W_{II} = \frac{240^2}{188(240-188)} = 5.892$$

The **total variation** is:

$$S_T = \text{Number of measurements} \times \text{number of analyzed classes}$$
$$= 12 \times 20 \times 2 = 480.$$

Variation of factor A, S_A, is obtained by weighting the level sums of each class as follows:

$$S_A = \left\{\frac{1}{240}(84-46)^2 \times W_I + \frac{1}{240}(108-80)^2 \times W_{II}\right\}$$

$$= 24.235 + 19.247$$

$$= 43.482$$

The **number of degrees of freedom** is (the number of levels – 1) × the number of analyzed classes:

$$f_A = (2-1) \times 2 = 2.$$

Variance, V_A, is the following:

$$V_A = \frac{S_A}{f_A} = \frac{43.482}{2} = 21.741$$

Because V_B and V_E are smaller than V_{e_2}, a **new error variance**, V_e, is obtained by pooling the variation of factors with small variance and repetition error variation S_{e_2}:

$$V_e = \frac{S_{e_2} + S_B + S_C + S_E + S_F + S_I + S_K}{f_{e_2} + f_B + f_C + f_E + f_F + f_I + f_K} = \frac{332.780}{468} = 0.711$$

The **variance ratio**, F_o, is obtained by dividing factor variance, V, by error variance, V_e.

$$F_{0A} = \frac{V_A}{V_e} = \frac{21.741}{0.711} = 30.58$$

The **contribution ratio** is calculated as follows:

$$\rho_A = \frac{S_A - f_A \times V_e}{S_T} \times 100 = \frac{43.482 - 2 \times 0.711}{480} \times 100 = 8.76\,(\%)$$

As for the **significance test**, factors are marked with ** if their variance ratio is greater than $F^2_{400}(0.01) = 4.66$. In this example, all significant factors are significant at the 1% level. Because the F value of F^2_{468} is not listed in the F Table, the F value with a smaller denominator (degrees of freedom), i.e., F^2_{400}, is used in this example. These results are summarized in *Table 8-19*.

Table 8-19: ANOVA Table—Burr Generation

Factor	f	S	V	F_o	Test	ρ(%)
A	2	43.482	21.741	30.58	**	8.76
B	2	0.269	0.134			
C	2	2.071	1.036			
D	2	12.251	6.125	8.61	**	2.26
E	2	1.172	0.586			
F	2	6.490	3.245			
G	2	19.917	9.958	14.00	**	3.85
H	2	40.994	20.497	28.83	**	8.24
I	2	2.810	1.405			
J	2	30.577	15.288	21.50	**	6.07
K	2	4.689	2.345			
e_2	456	315.279	0.691			
T	478	480.000				100.00
(e)	(468)	(332.780)	(0.711)			(70.81)

The calculation of the confidence limits for the process average can be omitted in the accumulation technique. If necessary, the calculation method explained in the previous section can be used.

8-7-2. Analysis of Variance for Research on Improvement of Picture-Image Formation—Film Adhesiveness and Picture Development (Corresponds to Subsection 6-2)

Analysis of Variance of Adhesiveness

Variation is calculated as follows:

Total variation:

$$S_T = (33.532)^2 + (31.707)^2 + \cdots + (29.018)^2 - \frac{(547.149)^2}{18}$$

$$= 48.635$$

The number of degrees of freedom is the number of data minus one, i.e., $f_T = 18 - 1 = 17$.

Variation, S_A, of factor A is calculated as follows:

$$S_A = \frac{1}{18}(272.84 - 274.31)^2$$

$$= \frac{1}{9}\{(272.84)^2 + (274.31)^2\} - \frac{(547.149)^2}{18}$$

$$= 0.119 \qquad (f_A = \text{number of levels} - 1 = 2 - 1 = 1)$$

Variation, S_B, is calculated as follows:

$$S_B = \frac{1}{6}\{(186.88)^2 + (186.95)^2 + (173.32)^2\} - \frac{(547.149)^2}{18}$$

$$= 20.535 \qquad (f_B = \text{number of levels} - 1 = 3 - 1 = 2)$$

The variation of other factors is similarly calculated.

Because the computer output of the level sum is given with only two digits below zero, the third digit below zero for variation may be different.

First order error variation, S_{e_1}, is obtained by subtracting all-factor variation from total variation (between-experiment variation), S_T.

$$S_{e_1} = S_T - (S_A + S_B + \cdots + S_H)$$

$$= 48.635 - (0.119 + 20.535 + \cdots + 3.801)$$

$$= 48.635 - 47.666$$

$$= 0.968 \qquad (f_{e_1} = 2)$$

The degrees of freedom, f_{e_1}, of first order error variance, S_{e_1} is:

$$f_{e_1} = f_T - (f_A + f_B + \cdots + f_H)$$

$$= 17 - (1 + 2 + \cdots + 2)$$

$$= 17 - 15$$

$$= 2$$

Variance is obtained by dividing variation by the degrees of freedom:

$$V_A = \frac{S_A}{f_A} = \frac{0.119}{1} = 0.119$$

$$V_B = \frac{S_B}{f_B} = \frac{20.535}{2} = 10.268$$

The rest are similarly calculated, and V_{e_1} is calculated as follows:

$$V_{e_1} = \frac{S_{e_1}}{f_{e_1}} = \frac{0.968}{2} = 0.484$$

New error variation, S_e, is calculated by pooling the variation of the factors with small variance, and a new error variance, V_{e_1}, is calculated from S_e and its number of degrees of freedom, f_e.

$$V_e = \frac{S_A + S_D + S_F + S_G + S_H + S_{e_1}}{f_A + f_D + f_F + f_G + f_H + f_{e_1}} = \frac{9.136}{11} = 0.831$$

Next, the **variance ratio** (F_o), **significance test**, and **contribution ratio** (ρ), are calculated for unpooled factors *B*, *C*, and *E*.

Variance ratios are as follows:

$$F_{oB} = \frac{10.268}{0.831} = 12.36$$

$$F_{oC} = \frac{6.334}{0.831} = 7.63$$

$$F_{oE} = \frac{3.147}{0.831} = 3.79$$

Because $F_{11}^2(0.01)$ is 7.20 and $F_{11}^2(0.05)$ is 3.98, factors with F_o greater than 7.20 are marked with **. Those between 7.20 and 3.98 are marked with *, and those less than 3.98 are marked with Δ.

The contribution ratio, ρ, is then calculated:

$$\rho = \frac{S - f \times V_e}{S_T} \times 100$$

$$\rho_B = \frac{20.535 - 2 \times 0.831}{48.635} \times 100 = 38.81\ (\%)$$

$$\rho_C = \frac{12.668 - 2 \times 0.831}{48.635} \times 100 = 22.63\ (\%)$$

$$\rho_E = \frac{6.295 - 2 \times 0.831}{48.635} \times 100 = 9.53\ (\%)$$

The **contribution ratio, ρ_e, of the error term** is $100 - (\rho_B + \rho_C + \rho_E) = 29.03$ (%).

Results of these calculations are summarized in *Table 8-20*.

The experiment itself can be judged to be a success because the contribution ratio of the error term is less than one-third of the total. The **confidence limit of the process average** of factors *B*, *D*, and *E* is calculated as follows:

$$\pm\sqrt{F_{fe}^1 \times \frac{V_e}{n_e}} = \pm\sqrt{4.84 \times \frac{0.831}{6}} = \pm 0.819$$

The **effective number of replications,** n_e, is given as:

$$n_e = \frac{18}{1 + f_B + f_C + f_E} = \frac{18}{7}$$

Table 8-20: ANOVA Table—Adhesiveness

Factor	f	S	V	F_0	Test	ρ(%)
A	1	0.119	0.119			
B	2	20.535	10.268	12.36	**	38.81
C	2	12.688	6.334	7.63	**	22.63
D	2	0.717	0.358			
E	2	6.295	3.147	3.79	Δ	9.53
F	2	1.209	0.604			
G	2	2.322	1.161			
H	2	3.801	1.900			
e_1	2	0.968	0.484			
T	17	48.635				100.00
(e)	(11)	(9.136)	(0.831)			(29.03)

The **confidence limit of, $\hat{\mu}$, the optimum condition,** is:

$$\pm\sqrt{F_{fe}^{1} \times \frac{V_e}{n_e}} = \pm\sqrt{4.84 \times \frac{0.831}{\left(\frac{18}{7}\right)}} = \pm 1.251$$

The **range of existence** of $\hat{\mu}$ is:

$$\hat{\mu}(B_1C_1E_3) = \overline{B}_1 + \overline{C}_1 + \overline{E}_3 - 2 \times \overline{T}$$
$$= 31.15 + 31.38 + 29.84 - 2 \times 30.397$$
$$= 31.576 \pm 1.251$$

Therefore, 30.325 ~ 32.827 [db].

Upper limit, m_u, and lower limit, m_l, are:

$$-10\log\frac{1}{m_l^2} = 30.325 \qquad m_l = 32.83$$

$$-10\log\frac{1}{m_u^2} = 32.827 \qquad m_u = 43.79$$

$\hat{\mu}$ is estimated to exist between 32.83 and 43.97, based on the observed (raw) data. The confirmatory trial results did not contradict this estimate.

Analysis of Variance of Picture Development

This can be conducted in the same manner as in the analysis of adhesiveness. *Table 8-21* is its ANOVA table. Readers are encouraged to try to obtain these calculation results by themselves.

Table 8-21: ANOVA Table—Picture Development

Factor	f	S	V	F_o	Test	ρ(%)
A	1	1.591	1.591			
B	2	88.053	44.026	30.61	**	64.61
C	2	2.725	1.362			
D	2	0.697	0.349			
E	2	9.781	4.891	3.40	Δ	5.24
F	2	4.271	2.136			
G	2	18.182	9.091	6.32	*	11.61
H	2	2.664	1.332			
e_1	2	3.873	1.936			
T	17	131.837				100.00
(e)	(11)	(15.822)	(1.438)			(18.55)

8-7-3. Analysis of Variance for Improvement Research on Chemical Plating Quantity Control and Variability of an Electronic Part, (Parameter Design, Corresponds to Section 6-3)

Analysis of Variance of Sensitivity, S

Variation is calculated as follows. (The level sum listed in *Table 5-10* is used):

$$\text{Total variation:}\quad S_T = (46.361)^2 + (43.434)^2 + \cdots + (54.971)^2 - \frac{(1694.27)^2}{36}$$

$$= 1101.97 \qquad (f_T = 36 - 1 = 35)$$

$$\text{Factor } A \text{ variation:}\quad S_A = \frac{1}{12}\left\{(569.54)^2 + (564.23)^2 + (560)^2\right\} - \frac{(1694.27)^2}{36}$$

$$= 3.484 \qquad (f_A = 3 - 1 = 2)$$

$$\text{Factor } B \text{ variation:}\quad S_B = \frac{1}{12}\left\{(565.27)^2 + (564.28)^2 + (564.72)^2\right\} - \frac{(1694.27)^2}{36}$$

$$= 0.023 \qquad (f_B = 3 - 1 = 2)$$

Variation for the other factors is similarly calculated.

First order error variance, S_{e_1} is obtained by subtracting the sum of assigned factor variations from total variation, S_T:

$$\text{Sum of factor variation} = S_A + S_B + S_C + \cdots + S_M$$

$$= 3.484 + 0.023 + 121.289 + \cdots + 5.102$$

$$= 1053.881$$

S_{e_1} and f_{e_1} are calculated as follows:

$$S_{e_1} = S_T - (\text{Sum of factor variation})$$
$$= S_T - (S_A + S_B + S_C + \cdots + S_M)$$
$$= 1107.97 - (3.484 + 0.023 + 121.289 + \cdots + 5.012)$$
$$= 1107.97 - 1053.88$$
$$= 48.083$$

$$f_{e_1} = f_T - (f_A + f_B + f_C + \cdots + f_M)$$
$$= 35 - (2 + 2 + 2 + \cdots + 2)$$
$$= 35 - (2 \times 13) = 35 - 26$$
$$= 9$$

Variance, V, is obtained by dividing variation by its degrees of freedom:

$$V_A = \frac{S_A}{f_A} = \frac{3.484}{2} = 1.742$$
$$V_B = \frac{S_B}{f_B} = \frac{0.023}{2} = 0.012$$
$$\vdots$$
$$V_{e_1} = \frac{S_{e_1}}{f_{e_1}} = \frac{48.083}{9} = 5.343$$

New error variation, S_e, is obtained by summing variation with small variance. If new error variation, S_e, is created by summing variation with variance less than 20, we obtain the following:

$$S_e = S_A + S_B + \cdots + S_M + S_{e_1}$$
$$= 3.484 + 0.023 + \cdots + 5.102 + 48.083$$
$$= 140.809$$

The **degrees of freedom of new error variance,** f_e, are then calculated:

$$f_e = f_A + f_B + \cdots + f_M + f_{e_1}$$
$$= 2 + 2 + \cdots + 2 + 9$$
$$= 27$$

New error variance, V_e, is calculated as follows:

$$V_e = \frac{S_e}{f_e} = \frac{140.809}{27}$$
$$= 5.215$$

The **variance ratio,** F_0, is obtained by dividing variance of unpooled factors, which were not added to create the new error variance, by the new error variance, V_e.

$$F_{0C} = \frac{V_C}{V_e} = \frac{60.645}{5.215}$$

$$= 11.63$$

$$F_{0D} = \frac{317.871}{5.215} = 60.95$$

$$F_{0E} = \frac{76.762}{5.215} = 14.72$$

$$F_{0I} = \frac{25.301}{5.215} = 4.85$$

The **significance test** is done, and we mark factors with ** if their variance ratio is greater than 5.49, with * if it is between 3.35 and 5.49, and with Δ if it is less than 3.35, i.e., $F^2_{27}(0.01) = 5.49$ and $F^2_{27}(0.05) = 3.35$.

The **contribution ratio,** ρ, is calculated only for marked factors:

$$\rho_C = \frac{S_C - f_C \times V_e}{S_T} \times 100 = \frac{121.289 - 2 \times 5.215}{1101.970} \times 100 = 10.06\ (\%)$$

$$\rho_D = \frac{635.742 - 2 \times 5.215}{1101.970} \times 100 = 56.75\ (\%)$$

$$\rho_E = \frac{153.523 - 2 \times 5.215}{1101.970} \times 100 = 12.99\ (\%)$$

$$\rho_I = \frac{50.602 - 2 \times 5.215}{1101.970} \times 100 = 3.65\ (\%)$$

The results of the above calculations are summarized in ANOVA *Table 8-22.*

The **confidence limit of the process average** is:

$$\pm\sqrt{F^1_{27} \times \frac{V_e}{n_e}} = \pm\sqrt{4.21 \times \frac{5.215}{12}} = \pm 1.353$$

The **effective number of replications,** n_e, is given as follows:

$$n_e = \frac{36}{1 + f_C + f_D + f_E + f_I} = \frac{36}{1+2+2+2+2} = 4.0$$

and **confidence limit of the optimum condition** is:

$$\pm\sqrt{F^1_{fe} \times \frac{V_e}{n_e}} = \pm\sqrt{4.21 \times \frac{5.215}{4}} = \pm 2.343$$

Analysis of Variance of the S/N Ratio, η

The same calculations as those used in the analysis of adhesiveness are used to create the ANOVA table, *Table 8-23.*

Table 8-22: Analysis of Variance Table—Senstivity (S_m)

Factor	f	S	V	F_o	Test	ρ(%)
A	2	3.484	1.742			
B	2	0.023	0.012			
C	2	121.289	60.645	11.63	**	10.06
D	2	635.742	317.871	60.95	**	56.75
E	2	153.523	76.762	14.72	**	12.99
F	2	35.070	17.535			
G	2	4.031	2.016			
H	2	9.344	4.672			
I	2	50.602	25.301	4.85	*	3.65
J	2	23.992	11.996			
K	2	2.414	1.207			
L	2	9.266	4.633			
M	2	5.102	2.551			
e_1	9	48.083	5.343			
T	35	1101.970				100.00
(e)	(27)	(140.809)	(5.215)			(16.56)

Table 8-23: ANOVA Table—S/N Ratio (η)

Factor	f	S	V	F_o	Test	ρ(%)
A	2	57.936	28.968	2.81	Δ	5.59
B	2	37.775	18.888	1.83	Δ	2.57
C	2	35.586	17.793	1.72	Δ	2.24
D	2	14.252	7.126			
E	2	53.953	26.977	2.61	Δ	5.00
F	2	36.303	18.151	1.76	Δ	2.35
G	2	0.834	0.417			
H	2	13.158	6.579			
I	2	92.045	46.023	4.46	*	10.71
J	2	62.850	31.425	3.04	Δ	6.33
K	2	44.705	22.353	2.17	Δ	3.61
L	2	69.953	34.977	3.39	Δ	7.40
M	2	20.533	10.267			
e_1	9	126.692	14.077			
T	35	666.575				100.00
(e)	(17)	(175.469)	(10.332)			(54.20)

Because the contribution of the error term is 54.20%, which is above 50%, it is expected to be difficult to control the S/N ratio, η . This should make us suspect that our parameter design study, which reduces variability through selection of the optimum combination, is insufficient. It may be necessary to control conditions within the limited range of condition values. In such a case, tolerance design should be carried out. The reader should consult Chapter 9 for details on tolerance design.

The confidence limit for the process average is:

$$\pm\sqrt{F_{fe}^{1} \times \frac{V_e}{n_e}} = \pm\sqrt{4.45 \times \frac{10.322}{12}} = \pm 1.956$$

The effective number of replications, n_e, is:

$$n_e = \frac{36}{1+(2\times 9)} = \frac{36}{19}$$

and confidence limit for the optimum condition is:

$$\pm\sqrt{F_{fe}^{1} \times \frac{V_e}{n_e}} = \pm\sqrt{4.45 \times \frac{10.322}{\left(\frac{36}{19}\right)}} = \pm 4.924$$

9

Analysis of Variance Applications: Tolerance Design and Process Control

This chapter introduces examples of the application of analysis of variance. In many cases one can easily locate the optimum condition without using analysis of variance. However, analysis of variance supplies more accurate information, and, in some areas, it is indispensable. This chapter explains tolerance design and the application of the Signal-to-Noise (S/N) ratio (for two kinds of error). These are the most important applications of analysis of variance in industrial experimental design.

9-1. Tolerance Design (Example: Electronic-Parts Chemical Plating-Process Tolerance Determination)

In this section, tolerance design is applied in the research on chemical plating uniformity improvement, which was discussed in Section 6-3.

With parameter design we determine the optimum level combination of control factors. If parameter design alone does not sufficiently reduce the impact of inner and outer noise, the variability caused by inner and outer noise factors must still be controlled, even if additional cost is incurred. The acceptable limits for a condition are called **tolerances** or acceptances. Tolerance design is a method for rationally determining tolerances.

9-1-1. Tolerance Design (Procedure and Analysis)

Step 1. Factor listing

Factors listed in a tolerance design study include the parts, circuit elements, and production processes that make up the system. Factors *A–M*, chosen for parameter design in Section 6-3, will be used here.

Steps 2–5. Level determination

Three levels are determined. The optimum condition is set as level two. Levels one and three are set as follows:

Level 1 $m - \sqrt{\frac{3}{2}}\ \sigma$

Level 2 m

Level 3 $m + \sqrt{\frac{3}{2}}\ \sigma$

The standard deviation of the level width of each factor is indicated by σ. Actual numbers for levels of factors *A–M* are summarized in *Table 9-1*.

Table 9-1: Factors and Levels

Symbol	Factor Name	Level 1	Level 2	Level 3
A	Drying temperature (°C)	76	80	84
B	Activation temperature (°C)	28	30	32
C	Composition ratio	0.75/1.25	0.8/1.2	0.85/1.15
D	Chemical plating temperature (°C)	75	77.5	80
E	Chemical plating density (%)	28	30	32
F	No. of rotations (rpm)	5	6	7
G	Oxygen processing time (sec)	8	10	12
H	Activating solution (g/l)	8	10	12
I	Activating time (min)	2.5	3	3.5
J	No. of shakings (rpm)	27	30	33
K	Exposure time	2.5	3	3.5
L	Flow quantity	1.25	1.50	1.75
M	Processing temperature	35	40	45

For example, the limit of ordinary deviation of factor *A*, for the least expensive equipment, is ±10°C. Thus, if we set 6σ = 20, then σ = 20/6.

Levels one and three are created by multiplying σ by $\pm\sqrt{3/2}$ and adding them to the central value.

Level 1 $80 - \sqrt{\frac{3}{2}}\sigma = 80 - \sqrt{\frac{3}{2}} \times \frac{20}{6} \cong 76$

Level 2 80

Level 3 $80 + \sqrt{\frac{3}{2}}\sigma = 80 + \sqrt{\frac{3}{2}} \times \frac{20}{6} \cong 84$

The levels of other factors are similarly determined.

Steps 6–9. Orthogonal array selection—experiment instructions

An L_{18} or L_{36} is used for tolerance design. Because 13 factors are assigned here, an L_{36} is used. *Table 9-2* shows the assignment. The experiment instructions table is omitted here. One point on each sample, from each experiment, is measured for chemical plating quantity. The experimental results are shown in the right hand column of *Table 9-2*. Tolerance design is conducted with the observed (raw) data.

Table 9-2: Layout and Data

Factor	A	B	C	D	E	F	G	H	I	J	K	L	M	Data
No.	1	2	3	4	5	6	7	8	9	10	11	12	13	
1	1	1	1	1	1	1	1	1	1	1	1	1	1	123
2	2	2	2	2	2	2	2	2	2	2	2	2	1	196
3	3	3	3	3	3	3	3	3	3	3	3	3	1	279
4	1	1	1	1	2	2	2	2	3	3	3	3	1	138
5	2	2	2	2	3	3	3	3	1	1	1	1	1	203
6	3	3	3	3	1	1	1	1	2	2	2	2	3	254
7	1	1	2	3	1	2	3	3	1	2	2	3	1	266
8	2	2	3	1	2	3	1	1	2	3	3	1	1	144
9	3	3	1	2	3	1	2	2	3	1	1	2	1	196
10	1	1	3	2	1	3	2	3	2	1	3	2	1	207
11	2	2	1	3	2	1	3	1	3	2	1	3	1	265
12	3	3	2	1	3	2	1	2	1	3	2	1	1	132
13	1	2	3	1	3	2	1	3	3	2	1	2	2	141
14	2	3	1	2	1	3	2	1	1	3	2	3	2	172
15	3	1	2	3	2	1	3	2	2	1	3	1	2	237
16	1	2	3	2	1	1	3	2	3	3	2	1	2	208
17	2	3	1	3	2	2	1	3	1	1	3	2	2	262
18	3	1	2	1	3	3	2	1	2	2	1	3	2	134
19	1	2	1	3	3	3	1	2	2	1	2	3	2	248
20	2	3	2	1	1	1	2	3	3	2	3	1	2	120
21	3	1	3	2	2	2	3	1	1	3	1	2	2	205
22	1	2	2	3	3	1	2	1	1	3	3	2	2	282
23	2	3	3	1	1	2	3	2	2	1	1	3	2	116
24	3	1	1	2	2	3	1	3	3	2	2	1	2	185
25	1	3	2	1	2	3	3	1	3	1	2	2	3	120
26	2	1	3	2	3	1	1	2	1	2	3	3	3	211
27	3	2	1	3	1	2	2	3	2	3	1	1	3	251
28	1	3	2	2	2	1	1	3	2	3	1	3	3	192
29	2	1	3	3	3	2	2	1	3	1	2	1	3	267
30	3	2	1	1	1	3	3	2	1	2	3	2	3	102
31	1	3	3	3	2	3	2	2	1	2	1	1	3	259
32	2	1	1	1	3	1	3	3	2	3	2	2	3	135
33	3	2	2	2	1	2	1	1	3	1	3	3	3	184
34	1	3	1	2	3	2	3	1	2	2	3	1	3	193
35	2	1	2	3	1	3	1	2	3	3	1	2	3	246
36	3	2	3	1	2	1	2	3	1	1	2	3	3	131

Steps 12–13. Level sum, average

Level sums are calculated by summing the data from experiments containing the particular level:

Level sum of A_1 $A_1 = 123 + 138 + \cdots + 193 = 2377$

Level sum of A_2 $A_2 = 196 + 203 + \cdots + 246 = 2337$

Level sum of A_3 $A_3 = 279 + 254 + \cdots + 231 = 2290$

The **average** is obtained by dividing the level sum by the number of data involved in calculating the level sum:

$$\overline{A}_1 = \frac{A_1}{12} = \frac{2377}{12} = 198.08$$

$$\overline{A}_2 = \frac{A_2}{12} = \frac{2337}{12} = 194.75$$

$$\overline{A}_3 = \frac{A_3}{12} = \frac{2290}{12} = 190.83$$

The level sum and the average of other factors are determined similarly. Calculation results for all factors are summarized in *Table 9-3*.

Table 9-3: Level Sum and Process Average

Lvl. / Factor	Level Sum			Average		
	1	2	3	1	2	3
A	2377.00	2337.00	2290.00	198.08	194.75	190.83
B	2354.00	2355.00	2295.00	196.17	196.25	191.25
C	2270.00	2312.00	2422.00	189.17	192.67	201.83
D	1536.00	2352.00	3116.00	128.00	196.00	259.67
E	2249.00	2334.00	2421.00	187.42	194.50	201.75
F	2354.00	2351.00	2299.00	196.17	195.92	191.58
G	2322.00	2353.00	2329.00	193.50	196.08	194.08
H	2343.00	2289.00	2372.00	195.25	190.75	197.67
I	2348.00	2307.00	2349.00	195.67	192.25	195.75
J	2294.00	2326.00	2384.00	191.17	193.83	198.67
K	2331.00	2314.00	2359.00	194.25	192.83	196.58
L	2322.00	2346.00	2336.00	193.50	195.50	194.67
M	2403.00	2310.00	2291.00	200.25	192.50	190.92
(*T*)	(7004)			(194.556)		

Step 14. Analysis of variance

The variation for each factor is calculated and divided into linear and quadratic variation, also called first and second order variation. Results are summarized in *Table 9-4*, and formulae are given on the next page.

Table 9-4: Decomposition of Variation

Level	$a = 2$	$a = 3$		$a = 4$			$a = 5$			
	b_1	b_1	b_2	b_1	b_2	b_3	b_1	b_2	b_3	b_4
W_1	−1	−1	1	−3	1	−1	−2	2	−1	1
W_2	1	0	−2	−1	−1	3	−1	−1	2	−4
W_3		1	1	1	−1	−3	0	−2	0	6
W_4				3	1	1	1	−1	−2	−4
W_5							2	2	1	1
$\lambda^2 S$	2	2	6	20	4	20	10	14	10	70
λS	1	2	2	10	4	6	10	14	12	24
S	1/2	2	2/3	5	4	9/5	10	14	72/5	288/35
λ	2	1	3	2	1	10/8	1	1	5/6	35/12

The first- and second-order coefficients are b_1 and b_2. A_1 and A_2 are the sum of r number of data. Coefficients W_1, W_2, ···, listed in column b_i, are used to derive the estimate of level a_ith order coefficient, b_i.

Estimation: $$b_i = \frac{W_1A_1 + W_2A_2 + \cdots + W_aA_a}{r(\lambda S)h^i} \qquad (i = 1, 2, \cdots)$$

Variation: $$S\,b_i = \frac{(W_1A_1 + W_2A_2 + \cdots + W_aA_a)^2}{r(\lambda^2 S)} \qquad (i = 1, 2, \cdots)$$

Number of units: $$b_i = \frac{1}{rSh^{2i}} \qquad (i = 1, 2\ \cdots)$$

Linear variation: $$S_l = \frac{(A_1 - A_3)^2}{12 \times 2} \qquad (f = 1)$$

Quadratic variation: $$S_q = \frac{(A_1 - 2A_2 + A_3)^2}{12 \times 6} \qquad (f = 1)$$

The sum of linear and quadratic variation, S_l and S_q, is equal to the original variation S:

S (original variation) = S_l (linear variation) + S_q (quadratic variation)

The equation for **linear variation,** S_l, is created by setting $W_1 = -1$, $W_2 = 0$, $W_3 = 1$, $\lambda^2 S = 2$, and $r = 12$. Furthermore, **quadratic variation,** S_q can be obtained by setting $W_1 = 1$, $W_2 = -2$, $W_3 = 1$, $\lambda^2 S = 6$, and $r = 12$. The number of degrees of freedom for each term of variation is one.

First, the **total variation,** S_T, is calculated by taking the squared sum of the difference between the data and the target value, 200:

$$S_T = (123 - 200)^2 + (196 - 200)^2 + \cdots + (131 - 200)^2$$

$$= 110160.00 \qquad f_T = 36$$

Next, the **variation, S_m**, of the difference between the target and the general mean is obtained:

$$\text{total sum} = 123 + 196 + \cdots + 131 = 7004$$

$$\text{target value } 36 \times 200 = 7200$$

$$S_m = \frac{(7004 - 7200)^2}{36} = \frac{(-196)^2}{36} = 1067.11$$

The **linear and quadratic variation of factor A**, S_{Al} and S_{Aq}, is calculated as follows:

$$S_{A\ell} = \frac{(A_1 - A_3)^2}{r(\lambda^2 S)} = \frac{(2377 - 2290)^2}{12 \times 2} = 315.38 \qquad (f = 1)$$

$$S_{Aq} = \frac{(A_1 - 2A_2 + A_3)^2}{r(\lambda^2 S)} = \frac{(2377 - 2 \times 2337 + 2290)^2}{12 \times 6} = 0.68 \qquad (f = 1)$$

Ordinary **undecomposed variation** is given as follows:

$$S_A = \frac{1}{12}(A_1^2 + A_2^2 + A_3^2) - \frac{(T)^2}{36} = \frac{1}{12}(2377^2 + 2337^2 + 2290^2) - \frac{(7004)^2}{36}$$

$$= 316.06 \qquad (f = 2)$$

Variation, S_A, is equal to the sum of linear and quadratic variation:

$$S_A = S_{A\ell} + S_{Aq} = 315.38 + 0.68 = 316.06$$

Variation, S_B, is similarly calculated.

$$S_{B\ell} = \frac{(2354 - 2295)^2}{12 \times 2} = 145.04 \qquad (f = 1)$$

$$S_{Bq} = \frac{(2354 - 2 \times 2355 + 2295)^2}{12 \times 6} = 51.68 \qquad (f = 1)$$

Variation, S_B, is equal to the sum of linear and quadratic variation:

$$S_B = \frac{1}{12}(2354^2 + 2355^2 + 2295^2) - \frac{(7004)^2}{36}$$

$$= 196.72 \qquad (f = 2)$$

The factor variation for other factors, C–M, is similarly calculated.

Primary error variation, S_{e_1}, is calculated by subtracting general variation, S_m, and the sum of factor variation from total variation, S_T.

Primary error variation is calculated as follows:

$$S_{e_1} = S_T - (S_m + S_{Al} + S_{Aq} + \cdots + S_{Mq})$$

$$= 110160 - (1067.11 + 315.38 + 0.68 + \cdots + 76.06)$$

$$= 615.51 \qquad f_{e_1} = 36 - (1 + 1 + \cdots + 1) = 9$$

Primary error variance, V_{e1}, is calculated as follows:

$$V_{e_1} = \frac{S_{e_1}}{f_{e_1}} = \frac{615.51}{9}$$

$$= 68.39$$

These results are summarized in *Table 9-5*, and a summary of the analysis of variance is shown in *Table 9-6*. A large linear term variation means that the factor level and the output are linearly related. A large quadratic term variation means that a strong nonlinear relationship exists between the the factor level and the output. Because quadratic term variation is an uncontrollable (noise) component, it is included in the error term. **Engineers need to control the variability of factors that have large linear term variance.**

Table 9-5: ANOVA Table

Factor		f	S	V	ρ (%)
m		1	1067.11	1067.11	**0.89
A	l	1	315.38	315.38	
	q	1	0.68	0.68	
B	l	1	145.04	145.04	
	q	1	51.68	51.68	
C	l	1	962.67	962.67	**0.79
	q	1	64.22	64.22	
D	l	1	104016.67	104016.67	**94.34
	q	1	37.56	37.56	
E	l	1	1232.67	1232.67	**1.04
	q	1	0.06	0.06	
F	l	1	126.04	126.04	
	q	1	33.35	33.35	
G	l	1	2.04	2.04	
	q	1	42.01	42.01	
H	l	1	35.04	35.04	
	q	1	260.68	260.68	
I	l	1	0.00	0.00	
	q	1	95.68	95.68	
J	l	1	337.50	337.50	

Table 9-5 Cont'd: Analysis of Variance Table

Factor		f	S	V	ρ (%)
	q	1	9.39	9.39	
K	l	1	32.67	32.67	
L	q	1	53.39	53.39	
	l	1	8.17	8.17	
M	q	1	16.06	16.06	
	l	1	522.67	522.67	
	q	1	76.06	76.06	
e_1		9	615.51	68.39	
T		36	110,160.00		100.00
(e)		(32)	(2880.88)	(90.0275)	(2.94)

Table 9-6: Rearranged ANOVA Table

Factor	f	S	V	ρ(%)
m	1	1067.11	1067.11	0.89
C_l	1	962.67	962.67	0.79
D_l	1	104016.67	104016.67	94.34
E_l	1	1232.67	1232.67	1.04
e	32	2880.88	90.0275	2.94
T	36	110,160.00		100.00

In our example, those factors are factors C, D, and E. The general mean, m, has a small but significant contribution. This is an effect due to the difference between the central value of the process and the target value. It was shown in the discussion of parameter design in Section 6-3 that factor D, which does not influence the variability, should be used for adjusting the mean. The output must be finely tuned to the target value using factor D.

Step 15. Drawing the factorial effects chart

The factorial-effects chart is drawn from the averages given in *Table* 9-3. Because factor D has a large and linear impact on the output, its level width must be limited. Factors with a larger linear effect have a larger linear term variation, S_l, and those with a nonlinear effect have a larger quadratic term variation, S_q. **Because the nonlinear component cannot be controlled by limiting the levels, it is treated as a noise component.**

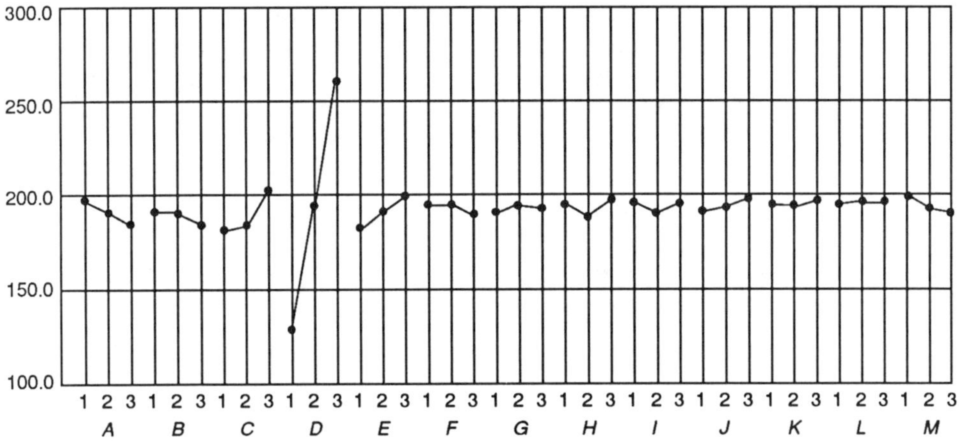

Figure 9-1: Factorial Effects Chart

Step 16. Factor tolerance determination

At this step, factors whose condition variability needs to be limited, compared to the current grade, are determined along with the extent of limitation. In our example, those factors are *C*, *D*, and *E*. Their first order variation was large in the analysis of variance. In particular, factor *D* has a strong effect, a 94.34% contribution ratio.

By controlling the condition variability of *D* so as to reduce the linear effect alone, the major part of the impact of condition variability on the performance can be attenuated. The general mean, *m*, is not adjusted in this example because factor *D* is sufficient to adjust the output to the target value.

The condition variability (tolerance) of factors *C*, *D*, and *E* is determined so as to minimize the loss after calculating the sum of the gain from quality improvement through process upgrading and the increase in cost due to equipment investment.

The gain from improved quality is calculated using the **Quality Loss Function**:

Quality Loss Function $L = \frac{A}{\Delta^2}\sigma^2$

where

A: Average loss caused by exceeding the tolerance (¥1,800)

Δ : Tolerance ($\pm$ 80)

σ^2: Variance

Variance, σ^2, is calculated as follows:

$$V_e = V_T\left[\left(\frac{1}{\alpha}\right)^2 \rho_C + \left(\frac{1}{\alpha}\right)^2 \rho_D + \left(\frac{1}{\alpha}\right)^2 \rho_E + \rho_e\right]$$

ρ_m is excluded because the target value (200) can be achieved by changing that level of factor *D*. α is the ratio of variability per grade when the current level is set at one.

Table 9-7 summarizes the total cost required to achieve the second, first, and super grades, which would reduce the condition variability ratio, n, to one-half, one-fifth, and one-tenth, respectively. The **total cost** includes initial investment, interest payments, and maintenance costs. Production quantity is assumed to be one million during the operation period.

Table 9-7: Grades and Total Cost

Grade	Variability (1/α)	C	D	E
Super	1/10	¥40 mil. (40)	¥30 mil. (30)	¥20 mil. (20)
First	1/5	¥20 mil. (20)	¥150 mil. (15)	¥10 mil. (10)
Second	1/2	¥5 mil. (5)	¥3 mil. (3)	¥2 mil. (2)
Current (parameter design)	1	Standard	Standard	Standard

( ): Cost increase per product

The following is used as the **estimate of average squared error,** σ^2:

$$V_T = \frac{1}{n}\left\{\sum (y_i - m)^2\right\}$$

where y_i is data, m is the target value, and n is the number of data.

In our example, the calculation is:

$$V_T = \frac{1}{36}\{S_T\} = \frac{1}{36}(110160.00) = 3060$$

The **grade combination** is estimated for several factors.

1. Current condition

$$L = \frac{1800}{80^2} \times 3060 = ¥860.63$$

2. Using second grade for all factors:

$$V_e = 3060\left[\left(\frac{1}{2}\right)^2 (0.0079 + 0.9434 + 0.0104) + 0.0294\right]$$

$$= 3060 \times 0.2698$$

$$= 825.66$$

The loss due to quality and cost increases will be as follows:

$$L = \frac{1800}{80^2} \times 825.66 + 5 + 3 + 2 = 232.22 + 10$$

$$= ¥242.22$$

3. Using first grade for factor D alone:

$$V_e = 3060\left[0.0079+\left(\frac{1}{5}\right)^2\times 0.9434+0.0104+0.0294\right]$$

$$= 3060\times 0.0854$$

$$= 261.43$$

The loss due to quality and cost increases will be as follows:

$$L=\frac{1800}{80^2}\times 261.43+15=73.53+15$$

$$= ¥88.53$$

4. Using first grade for all factors:

$$V_e = 3060\left[\left(\frac{1}{5}\right)^2\times 0.0079+\left(\frac{1}{5}\right)^2\times 0.9434+\left(\frac{1}{5}\right)^2\times 0.0104+0.0294\right]$$

$$= 3060\times 0.06787$$

$$= 207.68$$

The loss due to quality and cost increases will be as follows:

$$L=\frac{1800}{80^2}\times 207.68+20+15+10=58.40+45$$

$$= ¥103.41$$

5. Using super grade for factor D alone:

$$V_e = 3060\left[0.0079+\left(\frac{1}{10}\right)^2\times 0.9434+0.0104+0.0294\right]$$

$$= 3060\times 0.057134$$

$$= 174.83$$

The loss due to quality and cost increases will be as follows:

$$L=\frac{1800}{80^2}\times 174.83+30=49.17+30$$

$$= ¥79.17$$

6. Using super grade for all factors:

$$V_e = 3060\left[\left(\frac{1}{10}\right)^2(0.0079+0.9434+0.0104)+0.0294\right]$$

$$= 3060\times 0.03902$$

$$= 119.39$$

The loss due to quality and cost increases will be as follows:

$$L = \frac{1800}{80^2} \times 119.39 + 40 + 30 + 20 = 33.58 + 90$$
$$= ¥123.58$$

Loss (L) is expressed as the sum of the losses due to quality (Q) and cost increase (C).

L (loss) = quality (Q) + cost (C)

In general, quality improves as cost increases. Therefore, Q and C are inversely related. This relationship is shown in *Figure 9-2*. Curve L represents loss, Q is loss due to quality, and C is loss due to cost increase.

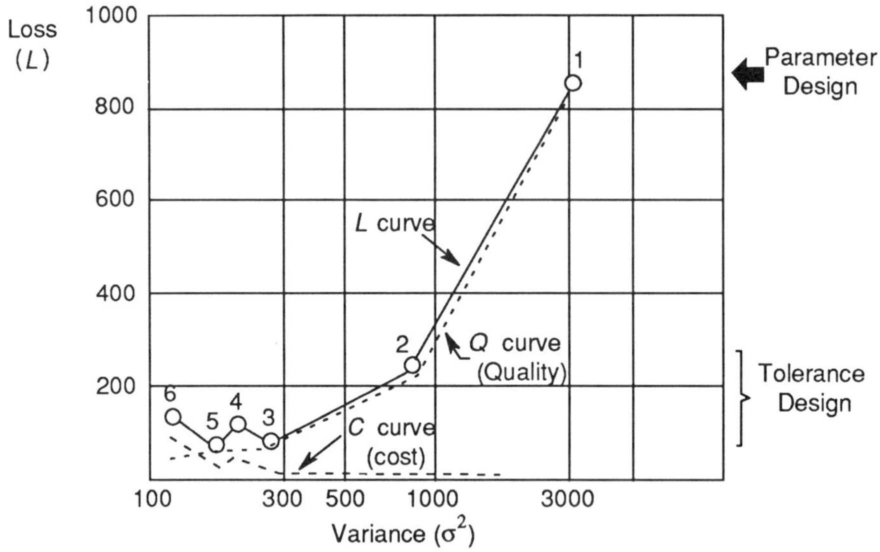

Figure 9-2: Relationship Between Loss and Variance

In our example, we need to choose the grade that gives the minimum value of L, i.e., (5).

The **gain due to tolerance design** will be:

$$L_{(1)} - L_{(5)} = 860.63 - 79.17$$
$$= ¥781.46 \quad \text{(per unit)}$$

Because annual production is one million, annual gain due to tolerance design will amount to:

$$781.46 \times 100 \text{ mil.} \cong ¥781 \text{ mil.}$$

This means that tolerance design created an additional gain of ¥781 million. The conclusion from tolerance design is that we should use only super grade for factor D.

The preceding calculation completes the tolerance design. The tolerance design experiment must be performed anticipating the use of the cheapest equipment and parts. It is recommended that information about inexpensive equipment and parts be obtained, even though their level width (or so-called accuracy) is small. As the grade of equipment and parts increases, they will become more expensive. Therefore, **the ideal should be to achieve a design at the parameter design stage that is robust against inner and outer noise with the output hitting the target value.**

9-1-2. Discussion of the Analysis

M. Did you understand the effectiveness of parameter design and tolerance design?

Q. I could follow the method and procedure, but I have never applied parameter design. Once the function is designed in the laboratory, it is immediately applied to the actual production machine.

M. The function design shown in *Table 9-8* has many levels.

Table 9-8: Function Design Levels and Product/Production Line Design

Mgt. Relation	Function Design Level	Product Design Content	Production Line Design Content
Kind of Product	New product development	• New system development • New production line design	• Trial run of the new production line • Condition determination • Final quality and function confirmation • Process adjustment
	Modified product	• Addition of new processes • Addition of options	• Condition determination of added processes • Apply existing condition to existing parts and processes
	Revised product	• Revision of process conditions • Revision of parts	• Revision of set value of the process condition
Quality	Production quantity increase	• Revision of the set value of each process condition	• Increase capability of existing equipment

Q. Although this table has some problems, I can understand it very well. I have not applied parameter design at any of the stages listed. The absolute capability and control capability required in a process are estimated on paper before undertaking a new production line design. The absolute capability is set high, and the control capability is limited to the severest level. Therefore, we do not consider how much the level combination of the process condition contributes to the central value of the characteristic value and to variability.

M. If that is the case, how do you determine the grade of the process?

Q. That is determined by the capability of the equipment sold in the marketplace when the new system is designed, and the cost of initial investment.

M. Then the process will give excess capacity.

Q. That is right. When parameter design is not conducted, the control capability of the process alone is used to control production line variability. This creates excess capacity from the viewpoint of parameter design engineers. Such a line often produces defective products, and when defects are produced and create customer claims, the designer will be blamed for having a process on the process control chart that lacks control capability.

M. When that happens, production line designers tend to use equipment with higher control capability and even use specially ordered parts, even if it costs more.

Q. Yes, that happens. By doing so, they will design a process without defects and claim losses.

M. But isn't it the case that increased initial investment raises the unit price of the product. Doesn't this result in a loss in market competitiveness on the part of the firm and extra cost to the consumers?

Q. A claim- and defect-free process is a good production process, but it generates a loss to the consumers. I think it is necessary to spend more to achieve better equipment.

M. But that is not beneficial to the consumers, even if the engineer in charge thinks it is good. From a different point of view, he is sacrificing the consumers for his own advantage. Furthermore, he is simply **imposing costs on consumers because of his lack of ability**. Even worse, he may go on to develop a good reputation within his company while he is actually committing multiple sins of stealing.

Q. That is true. I think parameter and tolerance design are better in the sense that they clearly define an effective investment segment with the smallest possible loss.

M. Why haven't engineers conducted design activities using parameter design before?

Q. I think because of two things about which engineers care the most. Their major concern is to produce acceptable products at the prototype production stage and to reduce defects and claims after the line has started production. Whether the product is acceptable or not is related to the central value, and the reduction of defects and claims is related to variability. Therefore, in order to produce good products with a small number of defects, engineers tend to create a system with a high control capability so as to control variability as narrowly as possible. Even novice engineers accept this idea after two or three failures.

M. That means that engineers with knowledge about the latest control equipment with high control capability are considered to be good engineers, is that right?

Q. Yes. In order to eliminate claims from the production line they have designed, engineers collect such information to defend themselves. The source of their information is catalogs. That is why they are called "catalog engineers."

M. What is the other reason?

Q. The other reason is that the relationship between objective characteristics of the product and the process is not clearly understood. We all know that development time is limited and engineers can obtain the conditions used to produce the prototype. Even in that case, they can best obtain one-way layout data with two levels of the main process condition.

M. What kind of information is necessary for line design?

Q. The factors of each process and their impact on the characteristic values. In addition, we want to know the impact of these factors on both the central value and variability.

What we want to know is which factor of which process must have control capability and to what extent. Design information about the existing process cannot be used to obtain such a determination. Even though we may not be able to determine anything, we still have to have the best design available at the design stage. **This forces us to produce a high-cost, excess-capacity production line**. In such a case, **production capacity is determined without regard to cost**. Therefore, it is impossible to even determine quality effectiveness.

M. This means that the amount of initial investment determines the upper limit of equipment capacity.

Q. That is right. The upper limit of process capability is determined by the highest possible amount of investment. In contrast, the methods of parameter design and tolerance design provide necessary and precise information.

M. For parameter design, it is necessary to perform experiments using orthogonal arrays with control factors assigned inside and error factors assigned outside. The following concept is very important. A prototype is nothing but proof of the fact that a product-fulfilling target function can be produced. There is no binding condition deriving from the way the prototype was created and tested. **The production process for the prototype must be used as a reference in repeated experiments to determine the combination of factor levels that fulfill the target function at minimum cost. This is the basic idea underlying parameter design.**

Q. Parameter design must be conducted regardless of whether we receive directions from the research and development group about the production method.

M. Parameter design is the only way of minimizing the loss to the company as a whole. Because parameter design can be done by the manufacturer, it must be considered a within-corporation activity. **Parameter design relates to the function of the software, and tolerance design relates to the function of the hardware.**

Parameter design does not increase costs in most cases.

Q. I understand that very well. I did not think there was any loss related to quality variability. I thought there would be no loss as long as the quality is within the standard.

M. In usual cases, the shipping tolerance is the limit used by companies in delivery inspection. Even if a product does not clear the tolerance by a large margin in company-conducted tests, it is not treated as a problem. However, consumers use products in various ways. Therefore, the loss is given by the following equation:

$$L = \frac{A}{(\Delta)^2} \times \sigma^2$$

where Δ is the tolerance, σ is variability, and A is the loss per product.

Q. Designers are interested in the initial investment, but they do not care about loss due to quality variability.

M. That is because producers pay the initial investment cost but it is the many unidentified consumers who pay the cost of quality variability. It is not acceptable for producers to plan unnecessary investments and secretly impose that cost on consumers.

Q. That is absolutely right. **Lazy engineers create loss to society.**

9-1-3. The Importance of Parameter Design Today

Q. I thought that the important things in production activities were product development capability and process control. But I changed my mind after studying parameter design.

M. Traditional design methods include only function design and tolerance design. In other words, **with traditional design methods the production system depends on equipment or system design and process control.** After World War II, quality control became popular and process control received attention, but the philosophy of design remained in its earlier mode. Today, the design requirements for corporate competition improvement are:

1. Start of production in a short period of time
2. Design of an inexpensive and highly reliable production line.

Q. **The idea today is to create a trouble-free line at the equipment function design stage**, isn't it? I think this is difficult to do, although it does not seem impossible.

M. I do not think it is impossible, either. Corporations that can do this will be the ones that will survive. The method for supporting this idea is being developed.

Process control is done to supplement deficiencies from the design stage. From this perspective, it is meaningless to perform extensive process control. Equipment should be designed so that a great deal of process control is not necessary. Otherwise, the design will be executed presupposing the necessity for massive process control. In today's context of small-quantity production of many kinds of products, the kinds of products may be changed before it is possible to collect the information necessary for process control. **Manufacturing today is not in an era of process control. Rather, it is in an era of design control.**

Q. Production departments today are also conducting uniform condition control through effective use of control charts.

M. Process control is an important production department activity. But it should not be done if it is not necessary. The traditional design method has another shortcoming, which is that the actual performance of a process cannot be known until one year has passed.

Q. Can you explain that?

M. It takes at least one year for a production process to experience outer noise caused by weather conditions, especially temperature and humidity changes that occur over four seasons. For example, a paper processing manufacturer experienced various paper cutting problems during the winter operation. These problems were caused by reduced tensile strength due to dryness in the winter when the absolute humidity is reduced.

Q. There are many cases in which performance varies depending on the season. In such a case, summer conditions or winter conditions are used.

M. If the characteristic value is different from the target value, the loss will be large. **The key point of design is to keep the characteristic value constant, even if there are changes in inner and outer noise.**

Q. I think that would be difficult with the traditional design method. The S/N ratio concept is indispensable.

M. On-line activities managed by production departments were the center of corporate activity before. But now, off-line activities, such as planning, development, and design, have moved to the center stage of corporate activity. **Corporate competition is now in the domain of upstream departments.** Planning and development belong to the realm of ideas and research, and design belongs to the realm of engineering technology, which minimizes the loss to society and maximizes corporate profits. It is the experimental design method and parameter design that contribute to engineering technology.

Q. What role does engineering technology play in management?

M. Any research that is not connected to management is meaningless. Even if a prototype is produced, it will not create profit. **Profit depends on engineering technology that produces a timely, robust product with a small cost.**

Q. I agree with you.

M. Engineers are inclined to be specialists who expect to become successful as independent, specialized engineers. But there is not much difference in the level of technological capability among engineers at different corporations.

Q. What should be done to realize the maximum potential of engineers?

M. The best and most effective method for realizing their engineering technology potential is to make them master the experimental design method.

Q. But this method is not yet accepted by engineers.

M. That is true—because they begin learning it by studying complicated subjects, such as analysis of variance. This bores them and the calculations cause them discomfort. "The problem lies in education" and, in this case, traditional education does not work.

Q. Why is that?

M. **As long as the experimental design method is taught as a mere set of techniques without a philosophy and context of application, engineers will not master it.** They consider it nothing more than common sense. It would be like studying English and then not going abroad—useless.

Q. What should be done?

M. Engineers are interested in contributing something to society through their own technology (ideas). All engineers will agree that they want to produce good products and supply them to society. **Engineers need to discuss what good products are.** Once the items of concern are clearly defined, they can start to discuss how to realize them. In this discussion, the item "timely creation of quality in design or process" will appear as the ideal condition. At this stage, the existence of an experimental design method which has been developed by many senior engineers should be introduced. By doing so, engineering technology can make a timely and secure contribution to management. This is a method that is necessary in order to bring the ideas of engineers to fruition (i.e., to help them realize their dreams). **Corporate technology is based on the know-how of corporate engineers and companies cannot compete without it.**

Q. Some companies have quite a low production-technology level compared to the level of individual engineering know-how.

M. Japanese management philosophy is highly advanced and is being studied by other countries. But the philosophy of engineering know-how remains undeveloped. The current level is limited to that of "tricks" and "crafts." **Engineers should have a philosophy about the relationship between society, the corporation, and technology.**

Q. A philosophy? I have never heard an engineer talk about his technology in relation to society.

M. That is right. In corporations, the corporate or presidential policy is presented once a year and is expressed in terms of management, technology, and competition.

Q. That is true at my company, too. What do the executives think that "technology" is?

M. I believe they think it is a superior technological discovery to be used at some opportune moment.

Q. If that.

M. Executives, managers, and engineers have different opinions about the importance of technology in corporate management policy. The latter tend to think in terms of engineering "technology." I think **a corporation's competitiveness depends on the maturity of its philosophy regarding technology.**

Q. Parameter design and tolerance design will promote a technology philosophy. Is that right?

M. I firmly believe that parameter design and tolerance design themselves constitute a great technology philosophy.

9-2. Experimental Design Method with Two Kinds of Error (Example: A Method for Determining the Optimum Condition of a Sensor in an Automation System)

Lately, sensor material, including organic sensor material, has been developing markedly. An automation system is composed of three subsystems: a sensor for signal input, a computer for signal processing, and a process control system. What is the method of determining the condition for deriving the maximum capability of an automation system, which is the hardware?

It is possible to select conditions such as the kind and position of a sensor, the power level of a sensor source, and the computer conditions that process a rather wide range of signals. The condition most advantageous for users must be chosen from among all possible conditions. Because that is the best condition, there must be only one.

If we have multiple conditions, we cannot say they are all best. Ideally, all combinations of conditions should be tested and confirmed for the determination of the best condition. But this is not practical because the number of experiments will be enormous. Therefore, it is necessary to have a method that rationally and practically assists in optimum condition selection, and that method is "the experimental design method." **Compared to the automation system, which is hardware, the method for rationally finding the optimum condition can be viewed as the software of the automation system.**

In order to make the material easy to understand, this section uses an automatic sorting machine as an example of an automation system. The automatic sorting machine was selected because it frees up workers from the most severe work of visual inspection, and reduction in the number of workers reduces labor costs.

9-2-1. Visual Inspection and the Sensor

The most important thing in replacing visual inspection with an automatic sorting machine is to obtain OK/NG judgments by the automatic sorting machine that are identical to good/inferior goods judgment made by visual inspection.

The results of visual inspection are expressed as good or inferior products and the results of judgment by the automatic sorting machine are expressed as OK or NG. The difference between visual inspection and inspection using the automatic sorting machine is that visual inspection covers fine points intuitively as it does total judgments, whereas the automatic sorting machine gives judgment based simply on the input signal.

Undesirable results (i.e., contradictory judgments between visual inspection and automatic sorting machine) will be generated for the failure types *A* and *B*, as shown in *Table 9-9*. Because visual inspection is the standard, the results of automatic sorting machine judgment must be adjusted to the results of visual inspection. Input signal strength and scanning condition are shown in *Figure 9-3* for failure types *A*, *B*, and *C* of *Table 9-9*.

Table 9-9: Failure Type and Difference in Judgment

Failure Type	Summary	Visual Inspection	Automatic Sorting Machine Judgment
A	Because the area of failure is small, this is not a problem.	Good (O)	NG
B	Although not a problem from a functional point of view, it is a problem because it is wide and large.	Inferior (X)	OK
C	Failure is large and strong. There are functional and appearance problems.	Inferior (X)	NG

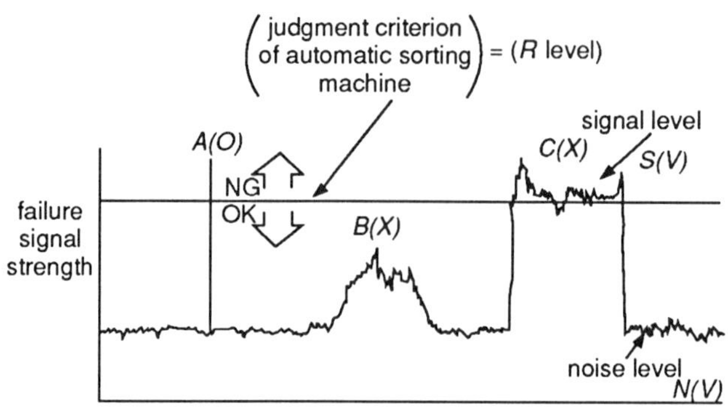

Figure 9-3: Scan Direction

In order for visual inspection results and automatic sorting machine judgments to coincide, the condition of the automatic sorting machine must be altered. The sensor section must be set so that the failure part is emphasized more than the noise level. This means that the condition should be adjusted to maximize the S/N ratio.

For the signal processing section, it is possible to trim the unnecessary portion of the input signal or carry out multiple judgments simultaneously. Optimization in the signal processing section is used to make the judgment of the automatic sorting machine coincide with the result of visual inspection. Here again, the condition is set so as to maximize the S/N ratio.

Therefore, **the automation system software sets the optimum condition of the automation system hardware (i.e., the automatic system itself).** The experimental design method is effective for this purpose.

The following introduces simple analytical procedures.

9-2-2. Discussion One: Two Kinds of Error

Q. I did not know that a rational method existed for deriving the maximum potential of an automation system.

M. I chose the automatic sorting machine example because it is the easiest one to understand. What is the most important thing for an automatic sorting machine?

Q. The most important thing is to make the automatic sorting machine judge those products that visual inspection would judge as good as OK and those products that visual inspection would judge to be inferior as NG. But it is sometimes difficult to classify products as good and inferior.

M. No loss will be generated if the automatic sorting machine judges products as NG or OK that cannot be judged good or inferior by visual inspection. There are **two kinds of errors**: judging a good product as NG and judging an inferior product as OK.

Q. Can you give an example?

M. Yes, I have a very interesting example.

A light-sensor system was used to count the number of attendees at the Science Exposition in Tsukuba, but the actual number (counted by the number of ticket stubs) differed from that counted by the sensor by 3,500 (at the entrance) and 7,800 (at the exit). The results are shown in *Table 9-10*.

We found several causes. The following problems were the most troublesome:

- When attendees passed through the gate as a tightly packed group, the sensor counted them as one person.
- When an umbrella passed through the gate, the sensor counted it as one person.

In order to cope with the second problem, two adjustment coefficients, for clear and rainy days, were prepared.

Table 9-10: Number of Attendees on Day 2 of the Tsukuba Science Exposition

No. of Ticket Stubs	**Sensor Count Number**	
Actual Numbers	Number at the Entrance	Number at the Exit
48,149	51,602	55,989
0	3,513	+7,840
—	0	+4,327

There were two kinds of errors: people not counted when they passed the gate as a group and counting umbrellas as a person. We need to set the automation system so that it counts people one by one and does not count umbrellas. This problem results from the fact that the hardware does not have an optimized system condition.

Q. What kind of examples exist on production sites, which are where an automation system is most frequently used?

M. An automatic sorting machine is a piece of equipment that replaces the visual inspection done by human beings, such as product appearance inspection.

There are two problems with an automatic sorting machine. The first is that defective products, which should have been sorted as NG, are delivered. This later generates claims. In addition, good products are sorted as NG. This causes the company's quality assurance section to order unnecessary improvements. The following example illustrates these problems.

9-2-3. Example of Sensor-System Development

1. Background

In order to reduce labor costs, a steel maker introduced an automatic sorting machine for sorting inspection in their color tin plate production. The automatic sorting machine is composed of a sensor, a marker, and a sorter. This section demonstrates the determination of the specification condition for the sensor, so as to accurately distinguish good and inferior products.

Twenty-five hundred test pieces were borrowed from the production section and were inspected at the inspection section. The samples were classified into 32 inferior products and 2,468 good products. The content of the complete sample inspection is summarized in *Table 9-11*. Data correspondence is shown in *Table 9-12*.

Table 9-11: Content or Sample Inspection for all 2,500 Test Pieces

Rank	Judgment	Number		Assigned Strength	Judgment Content
A	Good	2,436		Lowest	Both appearance and function are satisfactory
B	Good	20	2,468	Quite low	Ranking between *A* and *C*
C	Good	12		Low	Appearance relatively inferior
D	Inferior	10	32	High	Appearance inferior, but function is acceptable
E	Inferior	22		Quite high	Inferior function and appearance
Total		2,500	2,500		

Table 9-12: Correspondence Between Good/Inferior and OK/NG Judgments

I \ O	OK	NG	Total
Good Product	n_{00}	n_{01}	2,468
Inferior Product	n_{10}	n_{11}	32
Total	$n_{00}+n_{10}$	$n_{01}+n_{11}$	2,500

An orthogonal array was used to determine the sensor condition. The most desirable condition is for OK/NG judgments of the sensor to coincide with the inspection results. The judgment mechanism for OK/NG classification is shown in *Figure 9-4*.

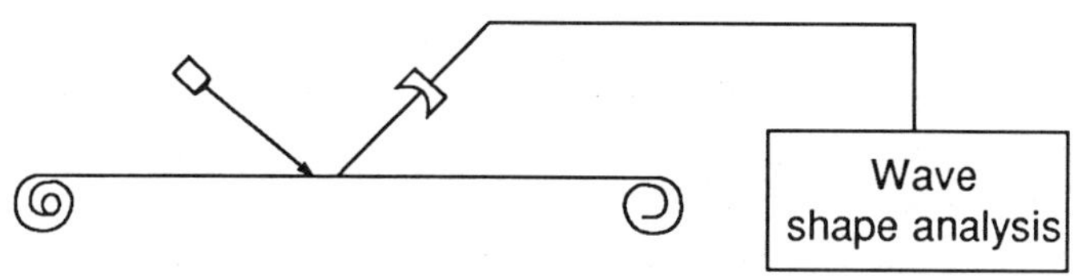

Figure 9-4: Sensor System Sketch

The output level of the sensor is *N* (v) if the sample is a good product. As for the inferior samples, the OK/NG judgment level, *R*, is set in between *S* (v). It is possible to change the OK/NG judgment level by means of *R*, and this is used to make the correction.

For each output of the corresponding orthogonal array experiment that determines the sensor condition, an OK/NG judgment was made at each R level, and data analysis was performed. The right hand column in *Table 9-12* is the data at $R = 1.25$.

2. Assignment and data analysis (Table 9-13)

   In *Table 9-13*, *A–C* are optical system factors and *D–K* are wave shape processing factors.

Table 9-13: Assignment and Raw Data at $R = 1.25$

Factor / Col. / No.	*A* (1 2 3)	*B* (4)	*C* (5)	*D* (6)	*E* (7)	*F* (8)	*G* (9)	*H* (10)	*I* (11)	*J* (12)	*K* (13)	*L* (14)	*M* (15)	Data n_{00}	n_{01}	n_{10}	n_{11}	Standard S/N Ratio [db]
1	1	1	1	1	1	1	1	1	1	1	1	1	1	2,443	25	3	29	8.56839
2	1	1	1	1	1	2	2	2	2	2	2	2	2	2,442	26	2	30	9.48917
3	1	2	2	2	2	1	1	1	1	2	2	2	2	2,446	22	5	27	7.50961
4	1	2	2	2	2	2	2	2	2	1	1	1	1	2,445	23	3	29	8.76327
5	2	1	1	1	2	1	1	1	2	1	1	2	2	2,444	24	4	28	7.91276
6	2	1	1	1	2	2	2	2	1	2	2	1	1	2,447	21	1	31	11.62255
7	2	2	2	2	1	1	1	1	2	2	2	1	1	2,444	24	4	28	7.91276
8	2	2	2	2	1	2	2	2	1	1	1	2	2	2,444	24	2	30	9.67417
9	3	1	2	1	2	1	2	1	2	1	2	1	2	2,445	23	6	26	6.88184
10	3	1	2	1	2	2	1	2	1	2	1	2	1	2,440	28	0	32	12.55924
11	3	2	1	2	1	1	2	1	2	2	1	2	1	2,449	19	10	22	5.71093
12	3	2	1	2	1	2	1	2	1	1	2	1	2	2,448	20	2	30	10.09365
13	4	1	2	2	1	1	2	2	1	1	2	2	1	2,451	17	6	26	7.60334
14	4	1	2	2	1	2	1	1	2	2	1	1	2	2,448	20	2	30	10.09335
15	4	2	1	1	2	1	2	2	1	2	1	1	2	2,449	19	3	29	9.20734
16	4	2	1	1	2	2	1	1	2	1	2	2	1	2,454	14	3	29	8.57309

3. Calculation of the standard S/N ratio

   The following is the example of $R = 1.25$ and experiment No.1 on the orthogonal array. *Table 9-14* shows the correspondence between the OK/NG signal when good and inferior products passed the sensor system. *Table 9-15* shows conversion of the results into the error ratio.

   The standard S/N ratio is calculated as follows:

   If the error ratio of the standard S/N ratio is P',

$$-10\log\left(\frac{1}{P'}-1\right)=\left\{-10\log\left(\frac{1}{0.010130}-1\right)-10\log\left(\frac{1}{0.09375}-1\right)\right\}\times\frac{1}{2}$$

$$=\frac{1}{2}\{-19.899834-9.85277\}$$

$$=14.8763$$

Table 9-14: Input/Output Number

I/O	OK	NG	Total
Good product	2, 443	25	2, 468
Inferior product	3	29	32
Total	2, 446	54	2, 500

Table 9-15: Input/Output Error Ratio

I/O	OK	NG	Total
Good product	0.98970	0.01030	1,000
Inferior product	0.09375	0.90625	1,000
Total	1.08345	0.91655	1,000

$P' = 0.031511$

$\rho = (1-2P')^2$

$\therefore$ Standard S / N ratio of No.1 $\eta_1 = -10\log(\frac{1}{\rho}-1)$

$$= 10\log(\frac{\rho}{1-\rho})$$

$$= 10\log\frac{(1-2P')^2}{1-(1-2P')^2}$$

$$= 10\log\frac{(1-2P')^2}{2P'\times(2-2P')}$$

$$= 10\log\frac{(1-2P')^2}{4P'(1-P')}$$

$$= 8.56839 \quad [\text{db}]$$

Standard S/N ratios for experiments No.2–No.16 were similarly calculated. When the error number is zero, the following is used:

the number of n_{10} is zero $\rightarrow q = 1/2(n_{10}+n_{11}) = 1/2\ T_1$

the number of n_{01} is zero $\rightarrow p = 1/2(n_{00}+n_{01}) = 1/2\ T_2$

where T_1 and T_2 are the total numbers.

The results are shown in *Table 9-16*.

Table 9-16: Standout S/N Ratio ($R = 1.25$)

No.	Standard S/N Ratio
1	8.56839
2	9.48917
3	7.50961
4	8.76327
5	7.91276
6	11.62255
7	7.91276
8	9.67417
9	6.88184
10	12.55924
11	5.71093
12	10.09365
13	7.60334
14	10.09365
15	9.20734
16	8.57309
	$T = 142.17576$

4. Analysis of Variance
   Level sums were calculated and are summarized in *Table 9-17*.

Table 9-17: Level Sum

Level / Factor	1	2	3	4
A	34.33044	37.12224	35.24564	35.47740
B	74.73094	67.44482		
C	71.17788	70.99788		
D	72.86600	69.30976		
E	69.14606	73.02970		
F	61.30697	80.86879		
G	73.22315	68.95261		
H	68.63423	73.54153		
I	76.83829	65.33747		
J	68.07051	74.10525		
K	72.48975	69.68601		
L	73.14345	69.03231		
M	71.31357	60.86219		

The variations were calculated as follows:

$$ST = (8.56839)^2 + (9.48917)^2 + \cdots + (8.57309)^2 - \text{CF} = 44.73206$$

$$T = 142.17576$$

$$\text{CF} = \frac{(142.17576)^2}{16} = 1263.37167$$

$$S_A = \frac{1}{4}(34.33044^2 + \cdots + 35.47740^2) - \text{CF} = 1.01426$$

$$S_B = \frac{1}{16}(74.73094 - 67.44482)^2 = 3.31796$$

Similarly,

$$S_C = 0.00203, \quad S_D = 0.79043, \quad S_E = 0.94267, \quad S_F = 23.91655,$$
$$S_G = 1.13984, \quad S_H = 1.50510, \quad S_I = 8.26680, \quad S_J = 2.27613,$$
$$S_K = 0.49131, \quad S_L = 1.05634, \quad S_M = 0.01273$$

These results are summarized in *Table 9-18.*

Table 9-18: ANOVA Table

Factor	f	S	V	F_0	ρ (%)
A	2	1.01426	—	—	—
B	1	3.31796	3.31796	6.07*	0.20
C	1	0.00203	—	—	—
D	1	0.79043	—	—	—
E	1	0.94267	—	—	—
F	1	23.91655	23.91655	43.89**	52.25
G	1	1.13984	—	—	—
H	1	1.50510	1.50510	2.76Δ	2.15
I	1	8.26680	8.26680	15.17**	17.26
J	1	2.27613	2.27613	4.18Δ	3.87
K	1	0.49131	—	—	—
L	1	1.05634	—	—	—
M	1	0.01273	—	—	—
T	15	44.73206			100.00
(e)	(10)	(5.44976)	(0.544976)		(18.27)

F_{10} = 10.00**
4.96*

The process averages of significant factors were calculated and are shown in *Figure 9-5*.

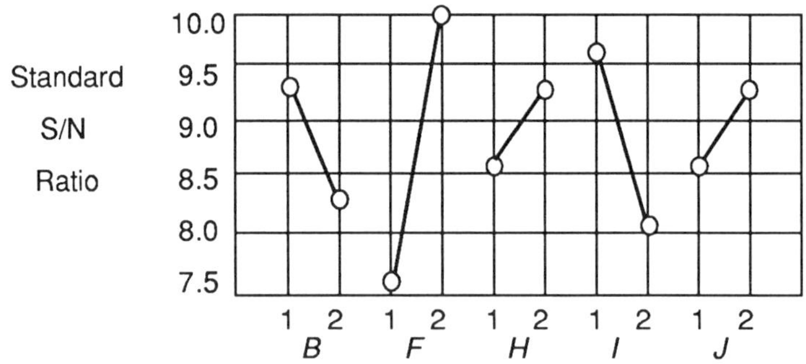

Figure 9-5: Factorial Effects

Next, the confidence limit was calculated:

$$\pm\sqrt{F \times V_e \times \frac{1}{r}} = \pm\sqrt{\frac{4.96 \times 0.544976}{8}} = \pm 0.581$$

5. Examination of the optimum condition

The levels of significant factors with high standard S/N ratios were selected as the optimum condition: B_1, F_2, H_2, I_1, and J_2.

$$\hat{\mu} = \overline{B}_1 + \overline{F}_2 + \overline{H}_2 + \overline{I}_1 + \overline{J}_2 - 4 \times \overline{T}$$
$$= 9.341 + 10.109 + 9.193 + 9.605 + 9.263 - 4 \times 8.896$$
$$= 11.967 \pm 1.007$$

$$\text{Confidence limit} = \pm\sqrt{\frac{4.96 \times 0.544976 \times 6}{10}} = \pm 1.007$$

The number of effective replications, n_e, is calculated as follows:

$$n_e = \frac{16 \text{ (total number of experiment)}}{f_m + f_B + f_F + f_H + f_I + f_J} = \frac{16}{6}$$

The contribution ratio, ρ, of the optimum condition was then estimated:

$$10 \log \frac{\rho}{1-\rho} = 11.967$$

$$\rho = 0.94022$$

The error ratio, P', when standardized, is estimated as follows (see *Table 9-19*):

Table 9-19: Standard Error Ratio

I/O	OK	NG	Total
Good product	0.98483	0.01517	1.0000
Inferior product	0.01517	0.98483	1.0000
Total	1.00000	1.00000	2.0000

$$(1 - 2P')^2 = 0.94022 \qquad P' = 0.01517$$

Therefore, the optimum condition of the hardware is $B_1, F_2, H_2, I_1,$ and J_2. The contribution ratio is $\rho = 0.94022$. The error ratio is $P = 0.01517$. In addition to the optimum condition of the hardware, the ratio of pq was adjusted so as to improve the total S/N ratio.

6. pq correction with adjustment factors

Adjustment of the pq ratio further improves the S/N ratio (software). The pq was adjusted proportionally to the inverse of their losses, A and B:

$$p:q = \frac{1}{A}:\frac{1}{B} \quad \frac{q}{A} = \frac{p}{B} \quad (\text{or } qB = pA)$$

Assume that the error ratio for judging a good product as NG is p, and for judging an inferior product as OK is q . The loss is assumed to be the following:

Loss for judging a good product as NG $A : A =$ ¥1,000 (p)
Loss for judging an inferior product as OK $B : B =$ ¥ 100 (q)

$$\frac{q}{1000} = \frac{p}{100} \qquad \therefore q = 10p$$

Substituting $q = 10p$ into the following equation and obtaining p and q after correction, we have:

$$\left(\frac{1}{P'} - 1\right)^2 = \left(\frac{10}{q} - 1\right) \times \left(\frac{1}{q} - 1\right)$$

By multiplying both sides by q and rearranging the equation, the following is derived:

$$\left(\frac{1}{P'} - 1\right)^2 \times q^2 = (10 - q) \times (1 - q) = q^2 - 11\,q + 10$$

$$\therefore \left[\left(\frac{1}{P'} - 1\right)^2 - 1\right] q^2 + 11\,q - 10 = 0$$

$$\therefore q = \frac{-11 \pm \left[11^2 - 4 \times \left\{\left(\frac{1}{P'} - 1\right)^2 - 1\right\} \times (-10)\right]^{\frac{1}{2}}}{2 \times \left\{\left(\frac{1}{P'} - 1\right)^2 - 1\right\}}$$

Since $R = 1.25$, $P' = 0.01513$

$$\therefore q = \frac{-11 \pm \left[11^2 - 4 \times \left\{\left(\frac{1}{0.01513} - 1\right)^2 - 1\right\} \times (-10)\right]^{1/2}}{2 \times \left\{\left(\frac{1}{0.01513} - 1\right)^2 - 1\right\}} = 0.047305$$

Because $q = 10\,p$, $p = 0.004731$.

The rest are calculated in the same way.

As a result, $q = 0.047428$ and $p = 0.004743$. These were then used to calculate the error ratio of the optimum condition after adjustment. The input/output data are shown in *Table 9-20* and *Table 9-21*.

Table 9-20: Error Ratio of Optimum Condition After Adjustment

I/O	OK	NG	Total
Good product	0.995257	0.004743	1.00000
Inferior product	0.04743	0.95257	1.00000
Total	1.042687	0.957313	2.00000

Table 9-21: Input/Output Table for Optimum Condition After Adjustment

I/O	OK	NG	Total
Good product	2,456.3	11.7	2,468
Inferior product	1.5	30.5	32
Total	2,457.8	42.2	2,500

The adjustment factor for p and q is R. The level of R must be determined so as to produce *Table 9-21*.

This concludes the determination of the optimum condition for the sensor, from the software aspect.

9-2-4. Discussion Two: Deriving of the Maximum Capability of the Equipment

Q. I do not understand the experimental design method itself.

M. First of all, I want you to understand that there is an effective approach to sensor system design that is different from traditional approaches.

Q. I understand that.

M. The use of experimental design is also spreading in the United States. In the beginning, chemical departments were the main users, but electronics, communications, and measurement departments have started using the method with new devices.

Q. I really want to study those.

M. There are many examples similar to the Tsukuba Science Exposition on/off counting example.

Q. That sounds interesting. Can you give me some?

M. Last year, there was a change in ¥1,000, ¥5,000, and ¥10,000 bills. Defective bills were in circulation even though every bill had been inspected by the bill sorting machine to eliminate defectives. The bill sorting machine committed two kinds of error, i.e., judging good ones as NG and judging defective ones as OK.

Q. Indeed, so even a top-of-the-line sorting machine had such a problem.

M. These two kinds of error are generated in more complicated systems. Another example involves automatic cola-vending machines. Sometimes they do not vend a cola when a ¥100 coin is inserted, and sometimes they deliver a can when the machine is kicked.

Q. I have had that experience, too. These kinds of things always happen.

M. Yes. Similar problems have been reported about mechatronics, automated machines, automatic sorting machines, and relays.

Q. What is the key point in sensor design?

M. As we have seen, the system must be designed so that a good product is judged OK and an inferior product is judged as NG. The experimental design method must be used effectively to derive the maximum capability of any system. The most common mistake committed by engineers is that they attempt to make the system perfect in the sense that all defects are judged NG.

In such a case, many good products are also judged NG, and that is wasteful. It is undesirable from the perspective of quality assurance and inspection.

Q. That is exactly right. I will try to master these methods.

The preceding section has discussed the method for determining the optimum condition for an automation system sensor.

Even when the latest automation system hardware is introduced, it will be meaningless if the maximum capability of the hardware itself in not utilized. The experimental design method is effective as an instrument for deriving its maximum capability. The experimental design method is not as difficult to master as its reputation would lead you to believe. The essence of the experimental method is becoming better understood in the United States. More examples of applications in the field of electronics, communications, and mechanical engineering have recently been accumulated in the U.S. than in Japan.

10

Computer Program

Manual calculation of level sums and process averages takes time and is sometimes inaccurate. There are several currently available computer programs that can do analysis of variance and draw factorial-effects charts. This chapter introduces a program for obtaining level sums and process averages and a program for Smaller-the-Better (STB), Larger-the-Better (LTB), and Nominal-the-Best (NTB) data transformation.

10-1. Programming for an L_{16}

This section describes a program for analyzing experiments with two repetitions. Although the language used here is BASIC, the content is applicable to any computer. In programming it is necessary to pay attention to the following points:

1. Make the number of assigned variables small (many variables may cause memory overflow)
2. Use a small initial value setting (recalculation may give different results)
3. Use simple statements (so that anyone can understand them)
4. Organize output statements (to make hand calculation easy).

Data shown in *Table 10-1* will be used to demonstrate analysis using the programs.

Table 10-1: L_{16} with Two Replications (Program Example)

Factor / Col. / No.	*A* 1	*B* 2	*C* 3	*D* 4	*E* 5	*F* 6	*G* 7	*H* 8	*I* 9	*J* 10	*K* 11	*L* 12	*M* 13	*N* 14	O 15	**Data**	
1	1	1	1	1	1	1	1	1	1	1	1	1	1	1	1	22.0	23.1
2	1	1	1	1	1	1	1	2	2	2	2	2	2	2	2	29.6	27.5
3	1	1	1	2	2	2	1	1	1	1	1	2	2	2	2	27.6	29.4
4	1	1	1	2	2	2	2	2	2	2	2	1	1	1	1	28.2	26.2
5	1	2	2	1	1	2	2	1	1	2	2	1	1	2	2	28.1	30.4
6	1	2	2	1	1	2	2	2	2	1	1	2	2	1	1	25.7	24.3
7	1	2	2	2	2	1	1	1	1	2	2	2	2	1	1	24.1	26.2
8	1	2	2	2	2	1	1	2	2	1	1	1	1	2	2	20.1	21.7
9	2	1	2	1	2	1	2	1	2	1	2	1	2	1	2	37.3	36.4
10	2	1	2	1	2	1	2	2	1	2	1	2	1	2	1	36.2	35.9
11	2	1	2	2	1	2	1	1	2	1	2	2	1	2	1	31.5	30.4
12	2	1	2	2	1	2	1	2	1	2	1	1	2	1	2	30.1	31.4
13	2	2	1	1	2	2	1	1	2	2	1	1	2	2	1	30.5	28.4
14	2	2	1	1	2	2	1	2	1	1	2	2	1	1	2	21.2	23.2
15	2	2	1	2	1	1	2	1	2	2	1	2	1	1	2	37.1	36.6
16	2	2	1	2	1	1	2	2	1	1	2	1	2	2	1	33.6	34.8

The following is the program flow:

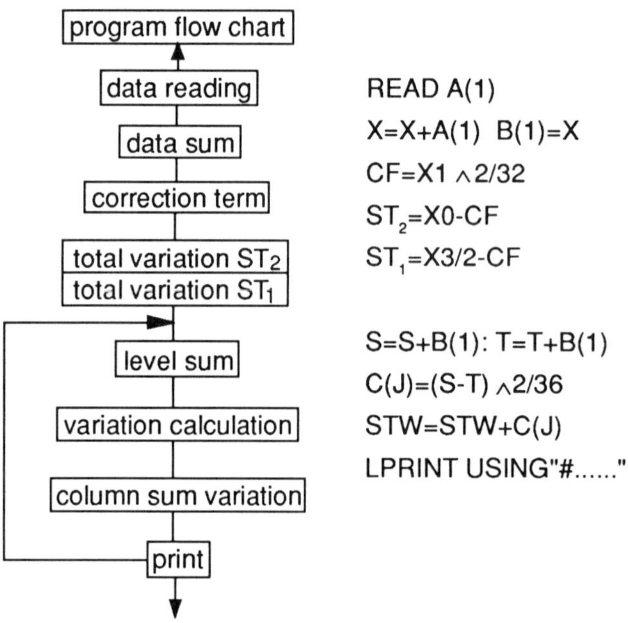

Actual program lists are given in the following: Rows 100–220 are total-variation print statements for obtaining between-experiment variation, S_{T_1}. $X1$ is the total sum, $X2$ is the squared sum of the data; and $X3$ is the square of the total sum of the data. The correction terms are $X1 \wedge 2 / 32$ and are contained in rows 205 and 210. Row 213 gives the error variance, S_{e_2}.

Variation Calculation Program

```
100 REM L16
110 CLEAR
120 DIM A(32), B(16), C(16)
125 X0=0: X1=0: X3=0
127 LPRINT "----NO----(1)----(2)----(WA)--"
130 FOR I=1 TO 16
135 LPRINT USING "#####"; I;
137 X=0: X2=0
140 FOR J=1 TO 2
150 READ A(I)
155 LPRINT USING"#####.#" ;A(I);
160 X=X+A(I): X2=X2+A(I) ^ 2
170 NEXT J
171 X1=X1+X :X3=X3+X ^ 2
172 LPRINT USING "#####.#" ; X
173 B(I)=X
180 X0=X0+X2
200 NEXT I
202 LPRINT "TOTAL ("X1")"
205 ST2=X0-X1 ^ 2/32
206 LPRINT "ST2("ST2")"
210 ST1=X3/2-X1 ^ 2/32
211 LPRINT "ST1("ST1")"
213 SE2=ST2-ST1
214 LPRINT "SE2("SE2")"
220 DATA 22.0, 23.1, 29.6, 27.5, 27.6, 29.4, 28.2, 26.2, 28.1, 30.4,
25.7, 24.3, 24.1, 26.2, 20.1, 21.7, 37.3, 36.4, 36.2, 35.9, 31.5, 30.4,
30.1, 31.4, 30.5, 28.4, 21.2, 23.2, 37.1, 36.6, 33.6, 34.8
290 LPRINT "---NO---(1)---(2)---(F)---(S)---(V)---(FO)---(R)---"
300 FOR J=1 TO 15
310 READ B$: S=0: T=0
320 FOR I=1TO 16
325 C$=MID$ (B$, I, 1)
330 IF C$="1" THEN S=S+B(I)   :GOTO 350
340 IF C$="2" THEN T=T+B(I)   :GOTO 350
350 NEXT I
360 C(J)=(S-T) ^ 2/32
405 LPRINT USING "#######";J;
410 LPRINT USING "######.#";S;
411 LPRINT USING "######.#";T;
420 LPRINT USING "###########.###";C(J)
```

Annotations:

- 100: comment statement (') non-executable statement
- 110: erase previous memory
- 120: array variable – declaration – yes: {integer: 32767; simple accuracy: 16383; double accuracy: 8191} "Subscript out of range"; no [upper limit is 10]
- 125: initial value setting
- 135: ";" – continuous display (print); "#####" – number of display letters specification
- 150: read data from data statement
- 160–171: X1 - total sum, X2 - Σ(each datum)^2; X3 - Σ(each experiment data)^2
- 172: opening
- 205: S_T; X1 ^ 2/32 – correction term
- 205–210: variation
- 210: S_{T1}
- 213: second order error Se_2
- 220: experiment data; DATA statement is better than INPUT statement for data input
- 290: output display specification
- 300: column correspondence (1~15)
- 310–320: experiment number correspondence (1~16)
- 330–340: level sum
- 360: column variation
- 405–420: print statement

```
430 STW=STW+C(J)        column variation sum
440 NEXT J
445 LPRINT "------------------------------------------------"
446 LPRINT
450 LPRINT "ST(SUM)("STW")"
510 DATA "111111122222222"
520 DATA "111222211112222"
530 DATA "111122222221111"
540 DATA "112211221122112"
550 DATA "112211222211221"
560 DATA "112222111122221"
570 DATA "112222112211122"
580 DATA "121212121212121"
590 DATA "121212122121212"
600 DATA "121221211212212"
610 DATA "121221212121121"
620 DATA "122112211221122"
630 DATA "122112212112211"
640 DATA "122121121221211"
650 DATA "122121122112122"
```

{ [ ] experiment points; the rest are program points }

L_{16} orthogonal array

Although we can input experimental results with input statements, the use of data statements that can be modified and moved freely is recommended. The experimental results are inputted in row 220 as a data statement.

Rows 290 to 650 give a program for obtaining the output table, level sums, and process averages. *J* and *I* correspond to the row number and the experiment number, respectively.

C$=MID$(B$, I, 1) is a command for reading the orthogonal array, which is inputted as character rows 510–650, word by word, to place it as C$. C$ is either 1 or 2. If it is "1," it is added to *S*. If it is "2," it is added to *T*. Therefore, S gives the level sum of the first level, and *T* gives the level sum of the second level. Row 360 squares the level sum difference and gives the variation *S* (*J*), by dividing it by the number of data (2 × 18 = 36). The sum of *S* (*J*) and S_{T_1} (between-experiment variation) must be the same. If they are different, check to see whether this is due to different numbers in the orthogonal array or due to errors at other places. The following are the programming checkpoints.

Programming Confirmation Points

1. Are input and output data the same? (There may be many input errors such as , and ,.)
2. Are the level sums of each column the same? (They will not be equal if there is an error in the orthogonal array.)
3. Are the between-experiment variation (S_{T_1}) and the column variation the same? (An error in initial value setting may cause this problem.)
4. Are data read correctly? (Is command RESTORE correctly placed?)
5. Is the result examined against the result of already known analysis of variance? (If the results are the same, the program is considered complete.)

10-2. Output Table for an L_{16} Array

Once program checking is done, the data and analysis of variance tables are obtained. They are shown in *Table 10-2* and *Table 10-3*. In *Table 10-2* the total sum (*T*), total variation (S_{T_2}), between-experiment variation (S_{T_1}), and second-order error variation (S_{E_2}) are shown together with each datum and data sum.

Table 10-2: Data Ouput Table

	Each Datum		Data Sum
NO	(1)	(2)	(WA)
1	22.0	23.1	45.1
2	29.6	27.5	57.1
3	27.6	29.4	57.0
4	28.2	26.2	54.4
5	28.1	30.4	58.5
6	25.7	24.3	50.0
7	24.1	26.2	50.3
8	20.1	21.7	41.8
9	37.3	36.4	73.7
10	36.2	35.9	72.1
11	31.5	30.4	61.9
12	30.1	31.4	61.5
13	30.5	28.4	58.9
14	21.2	23.2	44.4
15	37.1	36.6	73.7
16	33.6	34.8	68.4

Total (928.8) ← Total sum

S_{T_2} (810.64) ← Total variation

S_{T_1} (790.15) ← Between-experiment error

S_{e_2} (20.49) ← Second-order error variation

Table 10-3: ANOVA Table Output

Column Number	Level Sum		Degrees of Freedom	Variation	Variance (Manual Calculation)	Variance Ratio (Manual Calculation)	Contribution Ratio (Manual Calculation)
NO	(1)	(2)	(*F*)	(*S*)	(*V*)	(F_o)	(ρ)
1	414.2	514.6	1	315.005	315.005	185.77	38.65
2	482.8	446.0	1	42.320	42.320	24.96	5.01
3	459.0	469.8	1	3.645	—	—	—
4	459.8	469.0	1	2.645	—	—	—
5	476.2	452.6	1	17.405	17.405	10.26	1.94
6	482.2	446.6	1	39.605	39.605	23.36	4.68
7	421.0	507.8	1	235.445	235.445	138.85	28.84
8	479.1	449.7	1	27.011	27.001	15.93	3.12
9	457.3	471.5	1	6.301	—	—	—
10	442.3	486.5	1	61.051	61.051	36.00	7.32
11	460.1	468.7	1	2.311	—	—	—
12	462.3	466.5	1	0.551	—	—	—
13	451.9	476.9	1	19.531	19.531	11.52	2.20
14	453.1	475.7	1	15.961	15.961	9.41	1.76
15	461.1	467.7	1	1.361			
e_{22}			16	20.490	1.2806		
T			31	810.640			100.00
(*e*)			(22)	(37.304)	(1.6956)		(6.48)

In *Table 10-3*, the column numbers are taken vertically and level sums and averages are given. Variance, variance ratios, and contributions are manually calculated. It is necessary to confirm that the variation sum and S_{T_1} are equal. Manual calculation should be completed in about 15 minutes.

10-3. S/N Ratio Data-Transformation Programming

This section explains the Signal-to-Noise (S/N) ratio data transformation program used in Chapter 6. If data are continuous, they are either STB, LTB or NTB characteristic values.

10-3-1 S/N Ratio Transformation of NTB Data (L_{12})

Table 10-4 shows the program that transforms the data for hole-position variability improvement discussed in Chapter 6 into the S/N ratios. The command ERASE on row 215 erases memorized *N* (*J*).

Table 10-4 : S/N Ratio Transformation of NTB Data

```
110   CLEAR ' :L12 SN
120   DIM SN(12)
130   FOR I=1 TO 12 :DIM N(20):LPRINT USING "##" ;I; :X=0:X2=0
160   FOR J=1 TO 20
170   READ A :N(J)=A:LPRINT USING "###";A; :X=X+A
210   NEXT J:X1=X/20
212   XD=0:XE=0:XT=0
213   FOR J =1 TO 20
214   XD=N(J)–X1: XE=XE+XD^2 :XT=XT+XD
215   NEXT J:ERASE N
216   SN (I)=–10* (LOG(XE/20))/2.30259
220   LPRINT "T="; XT;
230   LPRINT USING "SN=####.###"; SN(I)
240   NEXT I
250   DATA –24, 10, 39, 4, 0, –23, 11, 22, 0, –19, 15, 14, 3, –19, 19, 21, 9, –5, –3, 38
260   DATA 21, 16, 3, 1, 25, 23, 8, –8, 2, 28, 1, 26, 30, 32, 4, 6, 20, 33, 19, 17
270   DATA 77, 35, 36, 67, 65, 47, 15, 48, 14, 48, 33, 17, 74, 57, 46, 66, 19, 36, 88, 7
280   DATA 42, 32, 19, 27, –64, 24, 25, 33, 69, 16, 13, 18, 24, 29, 12, 67, 17, –7, –1, 35
290   DATA 35, 30, 46, 29, 30, 0, 67, 32, 39, 20, 9, 37, 45, 45, 58, 28, 45, 24, 24, 9
300   DATA 68, 85, 87, 81, 31, 40, 47, 14, 16, 68, 67, 31, 31, 38, 53, 103, 60, 54, 98, 54
307   DATA –4, –13, 15, 21, –5, 21, –18, 2, 34, –4, –11, –5, 13, –14, 1, 13, –3, 5, –25, 32
308   DATA 38, 30, 66, 17, 6, 4, 23, 6, 30, 21, 36, 30, 36, 45, 0, 23, 9, 35, 15, 12
309   DATA 76, 49, 75, 76, –6, 26, 36, 54, 89, 134, 32, 60, 9, 34, 42, 64, 72, 14, 24, 13
310   DATA 8, –10, 12, 20, –2, 25, 14, 19, 6, 14, –4, 8, –7, 6, –6, –3, 16, 3, 13, 19
311   DATA 34, 17, 6, 31, 24, 12, 36, 0, 19, 9, 25, 26, 5, 0, 30, 13, 27, 0, 42, 20
312   DATA 13, –19, –18, –28, –19, –25, 0, –28, –10, –26, –21, 5, 14, –26, –10, –5, –28, –18, –17, –28
```

Data transformation uses the following formula:

$$\eta = -10\log\frac{1}{n}\left\{(y_1-\bar{y})^2+(y_2-\bar{y})^2+\cdots+(y_n-\bar{y})^2\right\}$$

10-3-2. S/N Ratio Transformation of STB and LTB Data (L_{18})

Table 10-5 and *Table 10-6* show the programs used for adhesiveness and picture developability of the picture formation film, discussed in Chapter 6. *I* indicates the experiment number of the L_{18}, and *J* is the number of data for each experiment.

Table 10-5: S/N Ratio Transformation of STB Data

```
110    CLEAR ' :L18MIN
120    DIM SN (18)
130    FOR I=1 TO 18 :LPRINT USING "## " ;I; :X=0:X2=0
160    FOR J=1 TO 2
170    READ A :A=A:LPRINT USING "###.##";A; :X=X+A :X2=X2+A^2
210    NEXT J :SN(I)=-(10*LOG(X2/2))/2.30259
220    LPRINT USING "###.##" ;X;
230    LPRINT USING "   SN=####.###" ; SN(I)
240    NEXT I
250    DATA 40, 43
260    DATA 22, 24
270    DATA 17, 19
274    DATA 39, 41
275    DATA 42, 45
276    DATA 46, 48
277    DATA 55, 58
278    DATA 49, 51
279    DATA 54, 56
280    DATA 29, 32
281    DATA 21, 33
282    DATA 38, 40
283    DATA 37, 41
284    DATA 50, 52
285    DATA 43, 45
286    DATA 55, 58
287    DATA 46, 47
288    DATA 52, 55
```

Table 10-6: S/N Ratio Transformation of LTB Data

```
110   CLEAR ' :L18MAX
120   DIM SN(18)
130   FOR I=1 TO 18 :LPRINT USING "## " ;I; :X=0:X2=0
160   FOR J=1 TO 2
170   READ A :A=A+21:LPRINT USING "###.## " ; A; :X=X+A :X2=X2+1/A^2
210   NEXT J :SN(I)=-(10*LOG(X2/2))/2.30259
220   LPRINT USING " ###.##";X;
230   LPRINT USING " SN=####.###" ; SN(I)
240   NEXT I
250   DATA    27,  26
260   DATA    18,  17
270   DATA     8,   6
274   DATA    24,  26
275   DATA    17,  15
276   DATA     8,   5
277   DATA     5,   4
278   DATA    11,  10
279   DATA     3,   2
280   DATA    18,  16
281   DATA    15,  10
282   DATA    14,  14
283   DATA    17,  16
284   DATA    16,  14
285   DATA    15,  14
286   DATA    13,  12
287   DATA     5,   5
288   DATA    15,   3
```

The data transformation follows the following formulae:

STB data $\eta = -10\log\frac{1}{n}\left(y_1^2 + y_2^2 + \cdots + y_n^2\right)$

LTB data $\eta = -10\log\frac{1}{n}\left(\frac{1}{y_1^2} + \frac{1}{y_2^2} + \cdots + \frac{1}{y_n^2}\right)$

10-3-3. Data Transformation for Parameter Design (L_{36})

Table 10-7 shows the data-transformation program for parameter design used in Section 6-3 for electronic parts chemical plating quantity. The program calculates both sensitivity and the S/N ratio.

$$\text{Sensitivity:}\quad S = 10\log\frac{1}{n}(y_1 + \cdots + y_n)^2$$

$$\text{S/N ratio:}\quad \eta = 10\log\frac{\frac{1}{n}(SM - V_e)}{V_e}$$

Because the recorded data are the result of subtracting 50 from each, 50 is added in row 170.

The variation calculation program of an L_{36}, transformed into the S/N ratio, is shown in Sub-section 8.7.3.

Table 10-7: Data Transformation Program for Parameter Design

```
110  CLEAR ' :L36 SM SN
120  DIM SM(36), SN(36)
130  FOR I=1 TO 36 :LPRINT USING "## " ;I; :X=0:X2=0
160  FOR J=1 TO 4
170  READ A :A=A+50 :LPRINT USING "### ";A; :X=X+A :X2=X2+A^2
210  NEXT J :SM(I)=(10*LOG(X^2/4))/2.30259 :VE=(X2-X^2/4)/3
215  SN(I)=+10*LOG((1/4*(X^2/4-VE)/VE)/2.30259
218  LPRINT USING "VE=###.###" ; VE;' :LPRINT USING " X^2#####.###" ;X2;
220  LPRINT USING " ###.##" ; X;
230  LPRINT USING " SM=####.###" ;SM(I); :LPRINT USING "  SN=###.###";SN(I)
240  NEXT I
250  DATA 58, 60, 59, 39
260  DATA 25, 22, 24, 26,
270  DATA 93, 89, 87, 90
274  DATA 213, 203, 208, 211
275  DATA 39, 34, 37, 44
276  DATA 23, 22, 22, 26
277  DATA 18, 12, 16, 20
278  DATA 368, 348, 365, 347
279  DATA 42, 42, 52, 43
280  DATA 54, 53, 44, 45
281  DATA 26, 22, 26, 31
282  DATA 174, 154, 141, 142
283  DATA 550, 549, 530, 528
284  DATA 8 , 10, 1, 3
285  DATA 31, 24, 28, 30
286  DATA 20, 14, 20, 17
287  DATA 36, 40, 28, 33
288  DATA 274, 277, 267, 261
289  DATA 64, 55, 53, 53
290  DATA 72, 86, 69, 71
291  DATA 40, 35, 35, 38
292  DATA 7, 5, 8, 11
293  DATA 164, 163, 158, 185
294  DATA 23, 32, 23, 24
295  DATA 378, 368, 437, 409
296  DATA 65, 66, 58, 60
297  DATA 4, 1, 4, 4
298  DATA 12, 14, 16, 14
299  DATA 12, 14, 16, 14
300  DATA 30, 28, 23, 18
301  DATA 33, 33, 18, 30
302  DATA 103, 105, 95, 81
303  DATA 23, 19, 20, 25
304  DATA 27, 26, 24, 23
305  DATA 3, 9, 0, 7
306  DATA 241, 224, 230, 226
```

Appendix

Appendix Table 1
F Table (5%, 1%)

Degrees of Freedom of the Population Variance of the Denominator	Degrees of Freedom of Variance of the Numerator										
f_2	1	2	3	4	5	6	7	8	9	10	11
1	161	200	126	225	230	234	237	239	241	242	243
	4 052	4 999	5 403	5 625	5 764	5 859	5 928	5 981	6 022	6 056	6 082
2	18.51	19.00	19.16	19.25	19.30	19.33	19.36	19.37	19.38	19.39	19.40
	98.49	99.01	99.17	99.25	99.30	99.33	99.34	99.36	99.38	99.40	99.41
3	10.13	9.55	9.28	9.12	9.01	8.94	8.88	8.84	8.81	8.78	8.76
	34.12	30.81	29.46	28.71	28.24	27.91	27.67	27.49	27.34	27.23	27.13
4	7.71	6.94	6.59	6.39	6.26	6.16	6.09	6.04	6.00	5.96	5.93
	21.20	18.00	16.69	15.98	15.52	15.21	14.98	14.80	14.66	14.54	14.45
5	6.61	5.79	5.41	5.19	5.05	4.95	4.88	4.82	4.78	4.74	4.70
	16.26	13.27	12.06	11.39	10.97	10.67	10.45	10.27	10.15	10.05	9.96
6	5.99	5.14	4.76	4.53	4.39	4.28	4.21	4.15	4.10	4.06	4.03
	13.74	10.92	9.78	9.15	8.75	8.47	8.26	8.10	7.98	7.87	7.79
7	5.59	4.74	4.35	4.12	3.97	3.87	3.79	3.73	3.68	3.63	3.60
	12.25	9.55	8.45	7.85	7.46	7.19	7.00	6.84	6.71	6.62	5.54
8	5.32	4.46	4.07	3.84	3.69	3.58	3.50	3.44	3.39	3.34	3.31
	11.26	8.65	7.59	7.01	6.63	6.37	6.19	6.03	5.91	5.82	5.74
9	5.12	4.26	3.86	3.63	3.48	3.37	3.29	3.23	3.18	3.13	3.10
	10.56	8.02	6.99	6.42	6.06	5.80	5.62	5.47	5.35	5.26	5.18
10	4.96	4.10	3.71	3.48	3.33	3.22	3.14	3.07	3.02	2.97	2.94
	10.04	7.56	6.55	5.99	5.64	5.39	5.21	5.06	4.95	4.85	4.78
11	4.84	3.98	3.59	3.36	3.20	3.09	3.01	2.95	2.90	2.86	2.82
	9.65	7.20	6.22	5.67	5.32	5.07	4.88	4.74	4.63	4.54	4.46
12	4.75	3.88	3.49	3.26	3.11	3.00	2.92	2.85	2.80	2.76	2.72
	9.33	6.93	5.95	5.41	5.06	4.82	4.65	4.50	4.39	4.30	4.22
13	4.67	3.80	3.41	3.18	3.02	2.92	2.84	2.77	2.72	2.67	2.63
	9.07	6.70	5.74	5.20	4.86	4.62	4.44	4.30	4.19	4.10	4.02
14	4.60	3.74	3.34	3.11	2.96	2.85	2.77	2.70	2.65	2.60	2.56
	8.86	6.51	5.56	5.03	4.69	4.46	4.28	4.14	4.03	3.94	3.86
15	4.54	3.68	3.29	3.06	2.90	2.79	2.70	2.64	2.59	2.55	2.51
	8.68	6.36	5.42	4.89	4.56	4.32	4.14	4.00	3.89	3.80	3.73
16	4.49	3.63	3.24	3.01	2.85	2.74	2.66	2.59	2.54	2.49	2.45
	8.53	6.23	5.29	4.77	4.44	4.20	4.03	3.89	3.78	3.69	3.61
17	4.45	3.59	3.20	2.96	2.81	2.70	2.62	2.55	2.50	2.45	2.41
	8.40	6.11	5.18	4.67	4.34	4.10	3.93	3.79	3.68	3.59	3.52
18	4.41	3.55	3.16	2.93	2.77	2.66	2.58	2.51	2.46	2.41	2.37
	8.28	6.01	5.09	4.58	4.25	4.01	3.85	3.71	3.60	3.51	3.44
19	4.38	3.52	3.13	2.90	2.74	2.63	2.55	2.48	2.43	2.38	2.34
	8.18	5.93	5.01	4.50	4.17	3.94	3.77	3.63	3.52	3.43	3.36
20	4.35	3.49	3.10	2.87	2.71	2.60	2.52	2.45	2.40	2.35	2.31
	8.10	5.85	4.94	4.43	4.10	3.87	3.71	3.56	3.45	3.37	3.30

Source: Genichi Taguchi, *Statistical Analysis*, Maruzen, 1972.

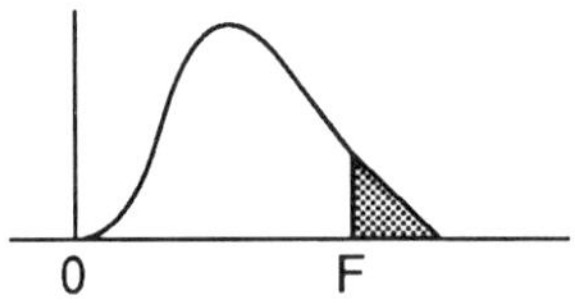

Top number is the 5% point of the F distribution
Bottom number is the 1% point of the F distribution

f_1

12	14	16	20	24	30	40	50	75	100	200	500	∞
244	245	246	248	249	250	251	252	253	253	254	254	254
6106	6142	6169	6208	6234	6258	6286	6302	6323	6334	6352	6361	6366
19.41	19.42	19.43	19.44	19.45	19.46	19.47	19.47	19.48	19.49	19.49	19.50	19.50
99.42	99.43	99.44	99.45	99.46	99.47	99.48	99.48	99.49	99.49	99.49	99.50	99.50
8.74	8.71	8.69	8.66	8.64	8.62	8.60	8.58	8.57	8.56	8.54	8.54	8.53
27.05	26.92	26.83	26.69	26.60	26.50	26.41	26.30	26.27	26.23	26.18	26.14	26.12
5.91	5.87	5.84	5.80	5.77	5.74	5.71	5.70	5.68	5.66	5.65	5.64	5.63
14.37	14.24	14.15	14.02	13.93	13.83	13.74	13.69	13.61	13.57	13.52	13.48	13.46
4.68	4.64	4.60	4.56	4.53	4.50	4.46	4.44	4.42	4.40	4.38	4.37	4.36
9.89	9.77	9.68	9.55	9.47	9.38	9.29	9.24	9.17	9.13	9.07	9.04	9.02
4.00	3.96	3.92	3.87	3.84	3.81	3.77	3.75	3.72	3.71	3.69	3.68	3.67
7.72	7.60	7.52	7.39	7.31	7.23	7.14	7.09	7.02	6.99	6.94	6.90	6.88
3.57	3.52	3.49	3.44	3.41	3.38	3.34	3.32	3.29	3.28	3.25	3.24	3.23
6.47	6.35	6.27	6.15	6.07	5.98	5.90	5.85	5.78	5.75	5.70	5.67	5.65
3.28	3.23	3.20	3.15	3.12	3.08	3.05	3.03	3.00	2.98	2.96	2.94	2.93
5.67	5.56	5.48	5.36	5.28	5.20	5.11	5.06	5.00	4.96	4.91	4.88	4.86
3.07	3.02	2.98	2.93	2.90	2.86	2.82	2.80	2.77	2.76	2.73	2.72	2.71
5.11	5.00	4.92	4.80	4.73	4.64	4.56	4.51	4.45	4.41	4.36	4.33	4.31
2.91	2.86	2.82	2.77	2.74	2.70	2.67	2.64	2.61	2.59	2.56	2.55	2.54
4.71	4.60	4.52	4.41	4.33	4.25	4.17	4.12	4.05	4.01	3.96	3.93	3.91
2.79	2.74	2.70	2.65	2.61	2.57	2.53	2.50	2.47	2.45	2.42	2.41	2.40
4.40	4.29	4.21	4.10	4.02	3.94	3.86	3.80	3.74	3.70	3.66	3.62	3.60
2.69	2.64	2.60	2.54	2.50	2.46	2.42	2.40	2.36	2.35	2.32	2.31	2.30
4.16	4.05	3.98	3.86	3.78	3.70	3.61	3.56	3.49	3.46	3.41	3.38	3.36
2.60	2.55	2.51	2.46	2.42	2.38	2.34	2.32	2.28	2.26	2.24	2.22	2.21
3.96	3.85	3.78	3.67	3.59	3.51	3.42	3.37	3.30	3.27	3.21	3.18	3.16
2.53	2.48	2.44	2.39	2.35	2.31	2.27	2.24	2.21	2.19	2.16	2.14	2.13
3.80	3.70	3.62	3.51	3.43	3.34	3.26	3.21	3.14	3.11	3.06	3.02	3.00
2.48	2.43	2.39	2.33	2.29	2.25	2.21	2.18	2.15	2.12	2.10	2.08	2.07
3.67	3.56	3.48	3.36	3.29	3.20	3.12	3.07	3.00	2.97	2.92	2.89	2.87
2.42	2.37	2.33	2.28	2.24	2.20	2.16	2.13	2.09	2.07	2.04	2.02	2.01
3.55	3.45	3.37	3.26	3.18	3.10	3.01	2.96	2.89	2.86	2.80	2.77	2.75
2.38	2.33	2.29	2.23	2.19	2.15	2.11	2.08	2.04	2.02	1.99	1.97	1.96
3.45	3.35	3.27	3.16	3.08	3.00	2.92	2.86	2.79	2.76	2.70	2.67	2.65
2.34	2.29	2.25	2.19	2.15	2.11	2.07	2.04	2.00	1.98	1.95	1.93	1.92
3.37	3.27	3.19	3.07	3.00	2.91	2.83	2.78	2.71	2.68	2.62	2.59	2.57
2.31	2.26	2.21	2.15	2.11	2.07	2.02	2.00	1.96	1.94	1.91	1.90	1.88
3.30	3.19	3.12	3.00	2.92	2.84	2.76	2.70	2.63	2.60	2.54	2.51	2.49
2.28	2.23	2.18	2.12	2.08	2.04	1.99	1.96	1.92	1.90	1.87	1.85	1.84
3.23	3.13	3.05	2.94	2.86	2.77	2.69	2.63	2.56	2.53	2.47	2.44	2.42

Degrees of Freedom of the Population Variance of the Denominator	Degrees of Freedom of Variance of the Numerator										
f_2	1	2	3	4	5	6	7	8	9	10	11
21	4.32	3.47	3.07	2.84	2.68	2.57	2.49	2.42	2.37	2.32	2.28
	8.02	5.78	4.87	4.37	4.04	3.81	3.65	3.51	3.40	3.31	3.24
22	4.30	3.44	3.05	2.82	2.66	2.55	2.47	2.40	2.35	2.30	2.26
	7.94	5.72	4.82	4.31	3.99	3.76	3.59	3.45	3.35	3.26	3.18
23	4.28	3.42	3.03	2.80	2.64	2.53	2.45	2.38	2.32	2.28	2.24
	7.88	5.66	4.76	4.26	3.94	3.71	3.54	3.41	3.30	3.21	3.14
24	4.26	3.40	3.01	2.78	2.62	2.51	2.43	2.36	2.30	2.26	2.22
	7.82	5.61	4.72	4.22	3.90	3.67	3.50	3.36	3.25	3.17	3.09
25	4.24	3.38	2.99	2.76	2.60	2.49	2.41	2.34	2.28	2.24	2.20
	7.77	5.57	4.68	4.18	3.86	3.63	3.46	3.32	3.21	3.13	3.05
26	4.22	3.37	2.89	2.74	2.59	2.47	2.39	2.32	2.27	2.22	2.18
	7.72	5.52	4.64	4.14	3.82	2.59	3.42	3.29	23.17	3.09	3.02
27	4.21	3.35	2.96	2.73	2.57	2.46	2.37	2.30	2.25	2.20	2.16
	7.68	5.49	4.60	4.11	3.79	3.56	3.39	3.26	3.14	3.06	2.98
28	4.20	3.34	2.95	2.71	2.56	2.44	2.36	2.29	2.24	2.19	2.15
	7.64	5.45	4.57	4.07	3.76	3.53	3.36	3.23	3.11	3.03	2.95
29	4.18	3.33	2.93	2.70	2.54	2.43	2.35	2.28	2.22	2.18	2.14
	7.60	5.52	4.54	4.04	3.73	3.50	3.33	3.20	3.03	3.00	2.92
30	4.17	3.32	2.92	2.69	2.53	2.42	2.34	2.27	2.21	2.16	2.12
	7.56	5.39	4.51	4.02	3.70	3.47	3.30	3.17	3.06	2.98	2.90
32	4.15	3.30	2.90	2.67	2.51	2.40	2.32	2.25	2.19	2.14	2.10
	7.50	5.34	4.46	3.97	3.66	3.42	3.25	3.12	3.01	2.94	2.86
34	4.31	3.28	2.88	2.65	2.49	2.38	2.30	2.23	2.17	2.12	2.08
	7.44	5.29	4.42	3.93	3.61	3.38	3.21	3.08	2.97	2.89	2.82
36	4.11	3.26	2.86	2.63	2.48	2.36	2.28	2.21	2.15	2.10	2.06
	7.39	5.25	4.38	3.89	3.58	3.35	3.18	3.04	2.94	2.86	2.78
38	4.10	3.25	2.85	2.62	2.46	2.35	2.26	2.19	2.14	2.09	2.05
	7.35	5.21	4.34	3.86	3.54	3.32	3.15	3.02	2.91	2.82	2.75
40	4.08	3.23	2.84	2.61	2.45	2.34	2.25	2.18	2.12	2.07	2.04
	7.31	5.18	4.31	3.83	3.51	3.29	3.12	2.99	2.88	2.80	2.73
42	4.07	3.22	2.83	2.59	2.44	2.32	2.24	2.17	2.11	2.06	2.02
	7.27	5.15	4.29	3.80	3.49	3.26	3.10	2.96	2.86	2.77	2.70
44	4.06	3.21	2.82	2.58	2.43	2.31	2.23	2.16	2.10	2.05	2.01
	7.24	5.12	4.26	3.78	3.46	3.24	3.07	2.94	2.84	2.75	2.68
46	4.05	3.20	2.81	2.57	2.42	2.39	2.22	2.15	2.09	2.04	2.00
	7.21	5.10	4.24	3.76	3.44	3.22	2.05	2.92	2.82	2.73	2.66
48	4.04	3.19	2.80	2.56	2.41	2.30	2.21	2.14	2.08	2.03	1.99
	7.19	5.08	4.22	3.74	3.42	3.20	3.04	2.90	2.80	2.71	2.64
50	4.03	3.18	2.79	2.56	2.40	2.29	2.20	2.13	2.07	2.02	1.98
	7.17	5.06	4.20	3.72	3.41	3.18	3.02	2.88	2.78	2.72	2.62
55	4.02	3.17	2.78	2.54	2.38	2.27	2.18	2.11	2.05	2.00	1.97
	7.12	5.01	4.16	3.68	3.37	3.15	2.98	2.85	2.75	2.66	2.59
60	4.00	3.15	2.76	2.52	2.37	2.25	2.17	2.10	2.04	1.99	1.95
	7.08	4.98	4.13	3.65	3.34	3.12	2.95	2.82	2.72	2.63	2.56
65	3.99	3.14	2.75	2.51	2.36	2.24	2.15	2.08	2.02	1.98	1.94
	7.04	4.95	4.10	3.62	3.31	3.09	2.93	2.79	2.70	2.61	2.54
70	3.98	3.13	2.74	2.50	2.35	2.32	2.14	2.07	2.01	1.97	1.93
	7.01	4.92	4.08	3.60	3.29	3.07	2.91	2.77	2.67	2.59	2.51

f_1

12	14	16	20	24	30	40	50	75	100	200	500	∞
2.25	2.20	2.15	2.09	2.05	2.00	1.96	1.93	1.89	1.87	1.84	1.82	1.81
3.17	3.07	2.99	2.88	2.80	2.72	2.63	2.58	2.51	2.47	2.42	2.38	2.36
2.23	2.18	2.13	2.07	2.03	1.98	1.93	1.91	1.87	1.84	1.81	1.80	1.78
3.12	30.2	2.94	2.83	2.75	2.67	2.58	2.53	2.46	2.42	2.37	2.33	2.31
2.20	2.14	2.10	2.04	2.00	1.96	1.91	1.88	1.84	1.82	1.79	1.77	1.76
3.07	2.97	2.89	2.78	2.70	2.62	2.53	2.48	2.41	2.37	2.32	2.28	2.26
2.18	2.13	2.09	2.02	1.98	1.94	1.89	1.86	1.82	1.80	1.76	1.74	1.73
3.03	2.93	2.85	2.74	2.66	2.58	2.49	2.44	2.36	2.33	2.27	2.23	2.21
2.16	2.11	2.06	2.00	1.96	1.92	1.87	1.84	1.80	1.77	1.74	1.72	1.71
2.99	2.89	2.81	2.70	2.62	2.54	2.45	2.40	2.32	2.29	2.23	2.19	2.17
2.15	2.10	2.05	1.99	1.95	1.90	1.85	1.82	1.78	1.76	1.72	1.70	1.69
2.96	2.86	2.77	2.66	2.58	2.50	2.41	2.36	2.28	2.25	2.19	2.15	2.13
2.13	2.08	2.03	1.97	1.93	1.88	1.84	1.80	1.76	1.74	1.71	1.68	1.67
2.93	2.83	2.74	2.63	2.55	2.47	2.38	2.33	2.25	2.21	2.16	2.12	2.10
2.12	2.06	2.02	1.96	1.91	1.87	1.81	1.78	1.75	1.72	1.69	1.67	1.65
2.90	2.80	2.71	2.60	2.52	2.44	2.35	2.30	2.22	2.18	2.13	2.09	2.06
2.10	2.05	2.00	1.94	1.90	1.85	1.80	1.77	1.73	1.71	1.68	1.65	1.64
2.87	2.77	2.68	2.57	2.49	2.41	2.32	2.27	2.19	2.15	2.10	2.06	2.03
2.09	2.04	1.99	1.93	1.89	1.84	1.79	1.76	1.72	1.69	1.66	1.64	1.62
2.84	2.74	2.66	2.55	2.47	2.38	2.29	2.24	2.16	2.13	2.07	2.03	2.01
2.07	2.02	1.97	1.91	1.86	1.82	1.76	1.74	1.69	1.67	1.64	1.61	1.59
2.80	2.70	2.62	2.51	2.42	2.34	2.25	2.20	2.12	2.08	2.02	1.98	1.96
2.05	2.00	1.95	1.89	1.84	1.80	1.74	1.71	1.67	1.64	1.61	1.59	1.57
2.76	2.66	2.58	2.47	2.38	2.30	2.21	2.15	2.08	2.04	1.98	1.94	1.91
2.03	1.98	1.93	1.87	1.82	1.78	1.72	1.69	1.65	1.62	1.59	1.56	1.55
2.72	2.62	2.54	2.43	2.35	2.26	2.17	2.12	2.04	2.00	1.94	1.90	1.87
2.02	1.96	1.92	1.85	1.80	1.76	1.71	1.67	1.63	1.60	1.57	1.54	1.53
2.69	2.59	2.51	2.40	2.32	2.22	2.14	2.08	2.00	1.97	1.90	1.86	1.84
2.00	1.95	1.90	1.84	1.79	1.74	1.69	1.66	1.61	1.59	1.55	1.53	1.51
2.66	2.56	2.49	2.37	2.29	2.20	2.11	2.05	1.97	1.94	1.88	1.84	1.81
1.99	1.94	1.89	1.82	1.78	1.73	1.68	1.62	1.60	1.57	1.54	1.51	1.49
2.64	2.54	2.46	2.35	2.26	2.17	2.08	2.02	1.94	1.91	1.85	1.80	1.78
1.98	1.92	1.88	1.81	1.76	1.72	1.66	1.63	1.58	1.56	1.52	1.50	1.48
2.62	2.52	2.44	2.32	2.24	2.15	2.06	2.00	1.92	1.88	1.82	1.78	1.75
1.97	1.91	1.87	1.80	1.75	1.71	1.65	1.62	1.57	1.54	1.51	1.48	1.46
2.60	2.50	2.42	2.30	2.22	2.13	2.04	1.98	1.90	1.86	1.80	1.76	1.72
1.96	1.90	1.86	1.79	1.74	1.70	1.64	1.61	1.56	1.53	1.50	1.47	1.45
2.58	2.48	2.40	2.28	2.20	2.11	2.02	1.96	1.88	1.84	1.78	1.73	1.70
1.95	1.90	1.85	1.78	1.74	1.69	1.63	1.60	1.55	1.52	1.48	1.46	1.44
2.56	2.46	2.39	2.26	2.18	2.10	2.00	1.94	1.86	1.82	1.76	1.71	1.68
1.93	1.88	1.83	1.76	1.72	1.67	1.61	1.58	1.52	1.50	1.46	1.43	1.41
2.53	2.43	2.35	2.23	2.15	2.06	1.96	1.90	1.82	1.78	1.71	1.66	1.64
1.92	1.86	1.81	1.75	1.70	1.65	1.59	1.56	1.50	1.48	1.44	1.41	1.39
2.50	2.40	2.32	2.20	2.12	2.03	1.93	1.87	1.79	1.74	1.68	1.63	1.60
1.90	1.85	1.80	1.73	1.68	1.63	1.57	1.54	1.49	1.46	1.42	1.39	1.37
2.47	2.37	2.30	2.18	2.09	2.00	1.90	1.84	1.76	1.71	1.64	1.60	1.56
1.89	1.84	1.79	1.72	1.67	1.62	1.56	1.53	1.47	1.45	1.40	1.37	1.35
2.45	2.35	2.28	2.15	2.07	1.98	1.88	1.82	1.74	1.68	1.62	1.56	1.53

Degrees of Freedom of the Population Variance of the Denominator f_2	Degrees of Freedom of Variance of the Numerator										
	1	2	3	4	5	6	7	8	9	10	11
80	3.96	3.11	2.72	2.48	2.33	2.21	2.12	2.05	1.99	1.95	1.91
	6.96	4.88	4.04	3.56	3.25	3.04	2.87	2.74	2.64	2.55	2.48
100	3.94	3.09	2.70	2.46	2.30	2.19	2.10	2.03	1.97	1.92	1.88
	6.90	4.82	3.98	3.51	3.20	2.99	2.82	2.69	2.59	2.51	2.43
125	3.92	3.07	2.68	2.44	2.29	2.17	2.08	2.01	1.95	1.90	1.86
	6.84	4.78	3.94	3.47	3.17	2.95	2.79	2.65	2.56	2.47	2.40
150	3.91	3.06	2.67	2.43	2.27	2.16	2.07	2.00	1.94	1.89	1.85
	6.81	4.75	3.91	3.44	3.13	2.92	2.76	2.62	2.53	2.44	2.37
200	3.89	3.04	2.65	2.41	2.26	2.14	2.05	1.98	1.92	1.87	1.83
	6.76	4.71	3.88	3.41	3.11	2.90	2.73	2.60	2.50	2.41	2.34
400	3.86	3.02	2.62	2.39	2.23	2.12	2.03	1.96	1.90	1.85	1.81
	6.70	4.66	3.83	3.36	3.06	2.85	2.69	2.55	2.46	2.37	2.29
1000	3.85	3.00	2.61	2.38	2.22	2.10	2.02	1.95	1.89	1.84	1.80
	6.66	4.62	3.80	3.34	3.04	2.82	2.66	2.53	2.43	2.34	2.26
∞	3.84	2.99	2.60	2.37	2.21	2.09	2.01	1.94	1.88	1.83	1.79
	6.64	4.60	3.78	3.32	3.02	2.80	2.64	2.51	2.41	2.32	2.24

f_1

	12	14	16	20	24	30	40	50	75	100	200	500	∞
	1.88	1.82	1.77	1.70	1.65	1.60	1.54	1.51	1.45	1.42	1.38	1.35	1.32
	2.41	2.32	2.24	2.11	2.03	1.94	1.84	1.78	1.70	1.65	1.57	1.52	1.49
	1.85	1.79	1.75	1.68	1.63	1.57	1.51	1.48	1.42	1.39	1.34	1.30	1.28
	2.36	2.26	2.19	2.06	1.98	1.89	1.79	1.73	1.64	1.59	1.51	1.46	1.43
	1.83	1.77	1.72	1.65	1.60	1.55	1.49	1.45	1.39	1.36	1.31	1.27	1.25
	2.33	2.23	2.15	2.03	1.94	1.85	1.75	1.68	1.59	1.54	1.46	1.40	1.37
	1.82	1.76	1.71	1.64	1.59	1.54	1.47	1.44	1.37	1.34	1.29	1.25	1.22
	2.30	2.20	2.12	2.00	1.91	1.83	1.72	1.66	1.56	1.51	1.43	1.37	1.33
	1.80	1.74	1.69	1.62	1.57	1.52	1.45	1.42	1.35	1.32	1.26	1.22	1.19
	2.28	2.17	2.09	1.97	1.88	1.79	1.69	1.62	1.53	1.48	1.39	1.33	1.28
	1.78	1.72	1.67	1.60	1.54	1.49	1.42	1.38	1.32	1.28	1.22	1.16	1.13
	2.23	2.12	2.04	1.92	1.84	1.74	1.64	1.57	1.47	1.42	1.32	1.24	1.19
	1.76	1.70	1.65	1.58	1.53	1.47	1.41	1.36	1.30	1.26	1.19	1.13	1.08
	2.20	2.09	2.01	1.89	1.81	1.71	1.61	1.54	1.44	1.38	1.28	1.19	1.11
	1.75	1.69	1.64	1.57	1.52	1.46	1.40	1.35	1.28	1.24	1.17	1.11	1.00
	2.18	2.07	1.99	1.87	1.79	1.69	1.59	1.52	1.41	1.36	1.25	1.15	1.00

Appendix Table 2 —Omega Transformation Table

p (%)	dB	p (%)	dB	p (%)	dB	p (%)	dB	p (%)	dB	p (%)	dB
0.0	∞*	5.0	-12.787	10.0	-9.541	15.0	-7.532	20.0	-6.020	25.0	-4.770
0.1	-29.995	5.1	-12.696	10.1	-9.493	15.1	-7.498	20.1	-5.993	25.1	-4.747
0.2	-26.980	5.2	-12.607	10.2	-9.446	15.2	-7.465	20.2	-5.966	25.2	-4.724
0.3	-25.215	5.3	-12.520	10.3	-9.399	15.3	-7.431	20.3	-5.939	25.3	-4.701
0.4	-23.961	5.4	-12.434	10.4	-9.352	15.4	-7.397	20.4	-5.912	25.4	-4.678
0.5	-22.988	5.5	-12.350	10.5	-9.305	15.5	-7.364	20.5	-5.885	25.5	-4.655
0.6	-22.191	5.6	-12.267	10.6	-9.259	15.6	-7.331	20.6	-5.859	25.6	-4.632
0.7	-21.518	5.7	-12.185	10.7	-9.214	15.7	-7.298	20.7	-5.832	25.7	-4.610
0.8	-20.933	5.8	-12.105	10.8	-9.168	15.8	-7.266	20.8	-5.806	25.8	-4.587
0.9	-20.417	5.9	-12.026	10.9	-9.124	15.9	-7.233	20.9	-5.779	25.9	-4.564
1.0	-19.955	6.0	-11.949	11.0	-9.079	16.0	-7.201	21.0	-5.753	26.0	-4.542
1.1	-19.537	6.1	-11.872	11.1	-9.035	16.1	-7.168	21.1	-5.727	26.1	-4.519
1.2	-19.155	6.2	-11.797	11.2	-8.991	16.2	-7.136	21.2	-5.701	26.2	-4.497
1.3	-18.803	6.3	-11.723	11.3	-8.947	16.3	-7.104	21.3	-5.675	26.3	-4.474
1.4	-18.476	6.4	-11.650	11.4	-8.904	16.4	-7.073	21.4	-5.649	26.4	-4.452
1.5	-18.172	6.5	-11.578	11.5	-8.861	16.5	-7.041	21.5	-5.623	26.5	-4.429
1.6	-17.888	6.6	-11.507	11.6	-8.819	16.6	-7.010	21.6	-5.598	26.6	-4.407
1.7	-17.620	6.7	-11.437	11.7	-8.777	16.7	-6.978	21.7	-5.572	26.7	-4.385
1.8	-17.367	6.8	-11.368	11.8	-8.735	16.8	-6.947	21.8	-5.547	26.8	-4.363
1.9	-17.128	6.9	-11.300	11.9	-8.693	16.9	-6.916	21.9	-5.521	26.9	-4.341
2.0	-16.901	7.0	-11.233	12.0	-8.652	17.0	-6.885	22.0	-5.496	27.0	-4.319
2.1	-16.685	7.1	-11.167	12.1	-8.611	17.1	-6.855	22.1	-5.470	27.1	-4.297
2.2	-16.478	7.2	-11.101	12.2	-8.570	17.2	-6.824	22.2	-5.445	27.2	-4.275
2.3	-16.281	7.3	-11.037	12.3	-8.530	17.3	-6.794	22.3	-5.420	27.3	-4.253
2.4	-16.091	7.4	-10.973	12.4	-8.490	17.4	-6.763	22.4	-5.395	27.4	-4.231
2.5	-15.910	7.5	-10.910	12.5	-8.450	17.5	-6.733	22.5	-5.370	27.5	-4.209
2.6	-15.735	7.6	-10.848	12.6	-8.410	17.6	-6.703	22.6	-5.345	27.6	-4.187
2.7	-15.566	7.7	-10.786	12.7	-8.371	17.7	-6.673	22.7	-5.321	27.7	-4.166
2.8	-15.404	7.8	-10.725	12.8	-8.332	17.8	-6.644	22.8	-5.296	27.8	-4.144
2.9	-15.247	7.9	-10.665	12.9	-8.293	17.9	-6.614	22.9	-5.271	27.9	-4.122
3.0	-15.096	8.0	-10.606	13.0	-8.255	18.0	-6.584	23.0	-5.247	28.0	-4.101
3.1	-14.949	8.1	-10.547	13.1	-8.216	18.1	-6.555	23.1	-5.222	28.1	-4.079
3.2	-14.806	8.2	-10.489	13.2	-8.178	18.2	-6.526	23.3	-5.198	28.2	-4.058
3.3	-14.668	8.3	-10.432	13.3	-8.141	18.3	-6.497	23.3	-5.1763	28.3	-4.036
3.4	-14.534	8.4	-10.375	13.4	-8.103	18.4	-6.468	23.4	-5.149	28.4	-4.015
3.5	-14.404	8.5	-10.319	13.5	-8.066	18.5	-6.439	23.5	-5.125	28.5	-3.994
3.6	-14.277	8.6	-10.263	13.6	-8.029	18.6	-6.410	23.6	-5.101	28.6	-3.972
3.7	-14.153	8.7	-10.209	13.7	-7.992	18.7	-6.381	23.7	-5.077	28.7	-3.951
3.8	-14.033	8.8	-10.154	13.8	-7.955	18.8	-6.353	23.8	-5.053	28.8	-3.930
3.9	-13.916	8.9	-10.100	13.9	-7.919	18.9	-6.325	23.9	-5.029	28.9	-3.909
4.0	-13.801	9.0	-10.047	14.0	-7.883	19.0	-6.296	24.0	-5.005	29.0	-3.888
4.1	-13.689	9.1	-9.994	14.1	-7.847	19.1	-6.268	24.1	-4.981	29.1	-3.867
4.2	-13.580	9.2	-9.942	14.2	-7.811	19.2	-6.240	24.2	-4.958	29.2	-3.846
4.3	-13.473	9.3	-9.890	14.3	-7.775	19.3	-6.212	24.3	-4.934	29.3	-3.825
4.4	-13.369	9.4	-9.839	14.4	-7.740	19.4	-6.184	24.4	-4.910	29.4	-3.804
4.5	-13.267	9.5	-9.788	14.5	-7.705	19.5	-6.157	24.5	-4.887	29.5	-3.783
4.6	-13.167	9.6	-9.738	14.6	-7.670	19.6	-6.129	24.6	-4.863	29.6	-3.762
4.7	-13.069	9.7	-9.688	14.7	-7.635	19.7	-6.101	24.7	-4.840	29.7	-3.741
4.8	-12.973	9.8	-9.639	14.8	-7.601	19.8	-6.074	24.8	-4.817	29.8	-3.720
4.9	-12.879	9.9	-9.590	14.9	-7.566	19.9	-6.047	24.9	-4.793	29.9	-3.699

*When p or r is less than 0.1, set n = 0/(1 – p) (or p/(1 – p) and use Appendix Table 4
Source: Shozo Konishi, et. al., *SN Ratio Manual*, Japanese Standard's Association, 1978

Appendix Table 2 (cont.) Omega Translation Table

P (%)	dB	P (%)	dB	P (%)	dB	P (%)	dB	P (%)	dB	P (%)	dB
30.0	–3.679	35.0	–2.687	40.0	–1.760	45.0	–0.871	50.0	0.000	55.0	0.872
30.1	–3.658	35.1	–2.668	40.1	–1.742	45.1	–0.853	50.1	0.017	55.1	0.889
30.2	–3.637	35.2	–2.649	40.2	–1.724	45.2	–0.835	50.2	0.035	55.2	0.907
30.3	–3.617	35.3	–2.630	40.3	–1.706	45.3	–0.818	50.3	0.052	55.3	0.924
30.4	–3.596	35.4	–2.611	40.4	–1.688	45.4	–0.800	50.4	0.069	55.4	0.942
30.5	–3.576	35.5	–2.592	40.5	–1.670	45.5	–0.783	50.5	0.087	55.5	0.959
30.6	–3.555	35.6	–2.573	40.6	–1.652	45.6	–0.765	50.6	0.104	55.6	0.977
30.7	–3.535	35.7	–2.554	40.7	–1.634	45.7	–0.748	50.7	0.122	55.7	0.995
30.8	–3.515	35.8	–2.536	40.8	–1.616	45.8	–0.730	50.8	0.139	55.8	1.012
30.9	–3.494	35.9	–2.517	40.9	–1.598	45.9	–0.713	50.9	0.156	55.9	1.030
31.0	–3.474	36.0	–2.498	41.0	–1.580	46.0	–0.695	51.0	0.174	56.0	1.047
31.1	–3.454	36.1	–2.479	41.1	–1.562	46.1	–0.678	51.1	0.191	56.1	1.065
31.2	–3.433	36.2	–2.460	41.2	–1.544	46.2	–0.660	51.2	0.209	56.2	1.083
31.3	–3.413	36.3	–2.441	41.3	–1.526	46.3	–0.643	51.3	0.226	56.3	1.100
31.4	–3.393	36.4	–2.423	41.4	–1.508	46.4	–0.625	51.4	0.243	56.4	1.118
31.5	–3.373	36.5	–2.404	41.5	–1.490	46.5	–0.608	51.5	0.261	56.5	1.136
31.6	–3.353	36.6	–2.385	41.6	–1.472	46.6	–0.591	51.6	0.278	56.6	1.153
31.7	–3.333	36.7	–2.366	41.7	–1.454	46.7	–0.573	51.7	0.295	56.7	1.171
31.8	–3.313	36.8	–2.348	41.8	–1.436	46.8	–0.556	51.8	0.313	56.8	1.189
31.9	–3.293	36.9	–2.329	41.9	–1.419	46.9	–0.538	51.9	0.330	56.9	1.206
32.0	–3.273	37.0	–2.310	42.0	–1.401	47.0	–0.521	52.0	0.348	57.0	1.224
32.1	–3.253	37.1	–2.292	42.1	–1.383	47.1	–0.503	52.1	0.365	57.1	1.242
32.3	–3.233	37.2	–2.273	42.2	–1.365	47.2	–0.486	52.2	0.382	57.2	1.260
32.3	–3.213	37.3	–2.255	42.3	–1.347	47.3	–0.468	52.3	0.400	57.3	1.277
32.4	–3.193	37.4	–2.236	42.4	–1.330	47.4	–0.451	52.4	0.417	57.4	1.295
32.5	–3.173	37.5	–2.217	42.5	–1.312	47.5	–0.434	52.5	0.435	57.5	1.313
32.6	–3.153	37.6	–2.199	42.6	–1.294	47.6	–0.416	52.6	0.452	57.6	1.331
32.7	–3.134	37.7	–2.180	42.7	–1.276	47.7	–0.399	52.7	0.469	57.7	1.348
32.8	–3.114	37.8	–2.162	42.8	–1.259	47.8	–0.381	52.8	0.487	57.8	1.366
32.9	–3.094	37.9	–2.144	42.9	–1.241	47.9	–0.364	52.9	0.504	57.9	1.384
33.0	–3.075	38.0	–2.125	43.0	–1.223	48.0	–0.347	53.0	0.522	58.0	1.402
33.1	–3.055	38.1	–2.107	43.1	–1.205	48.1	–0.329	53.1	0.539	58.1	1.420
33.2	–3.035	38.2	–2.088	43.2	–1.188	48.2	–0.312	53.2	0.557	58.2	1.437
33.3	–3.016	38.3	–2.070	43.3	–1.170	48.3	–0.294	53.3	0.574	58.3	1.455
33.4	–2.996	38.4	–2.051	43.4	–1.152	48.4	–0.277	53.4	0.592	58.4	1.473
33.5	–2.977	38.5	–2.033	43.5	–1.135	48.5	–0.260	53.5	0.609	58.5	1.491
33.6	–2.957	38.6	–2.015	43.6	-1.117	48.6	–0.242	53.6	0.626	58.6	1.509
33.7	–2.938	38.7	–1.996	43.7	–1.099	48.7	–0.225	53.7	0.644	58.7	1.527
33.8	–2.918	38.8	–1.978	43.8	–1.082	48.8	–0.208	53.8	0.661	58.8	1.545
33.9	–2.899	38.9	–1.960	43.9	–1.064	48.9	–0.190	53.9	0.679	58.9	1.563
34.0	–2.880	39.0	–1.942	44.0	–1.046	49.0	–0.173	54.0	0.696	59.0	1.581
34.1	–2.860	39.1	–1.923	44.1	–1.029	49.1	–0.155	54.1	0.714	59.1	1.599
34.2	–2.841	39.2	–1.905	44.2	–1.011	49.2	–0.138	54.2	0.731	59.2	1.617
34.3	–2.822	39.3	–1.887	44.3	–0.994	49.3	–0.121	54.3	0.749	59.3	1.635
34.3	–2.802	39.4	–1.869	44.4	–0.976	49.4	–0.103	54.4	0.766	59.4	1.653
34.5	–2.783	39.5	–1.851	44.5	–0.958	49.5	–0.086	54.5	0.784	59.5	1.671
34.6	–2.764	39.6	–1.832	44.6	–0.941	49.6	–0.068	54.6	0.801	59.6	1.689
34.7	–2.745	39.7	–1.814	44.7	–0.923	49.7	–0.051	54.7	0.819	59.7	1.707
34.8	–2.726	39.8	–1.796	44.8	–0.906	49.8	–0.034	54.8	0.836	59.8	1.725
34.9	–2.707	39.9	–1.778	44.9	–0.888	49.9	–0.016	54.9	0.854	59.9	1.743

Appendix Table 3
Problems Affecting Research and Development and Countermeasures

Problem	Situation	Countermeasure (way of thinking)
Importance is attached to theory, but it is not related to new discovery.	The theory does not go beyond common sense. There is no invention which leads to new product development.	A theory mathematically describes a phenomenon. Theory does not precede phenomenon. A theory describes a simple phenomenon. An invention is the utilization of complex phenomena. There is no theory for complex phenomena.
No leadership	Individual activity becomes the higher priority.	The role of a project leader is to support and to remove obstacles to engineers. The role of a leader is not to rule, suppress or manage. Project teams are normally disordered. It is more important to give a broad direction rather than doing trivial management.
Individual does not grow	The attempt to train individuals through projects does not work.	Explain and make people understand the meaning, the philosophy of corporate research and competition. Rather than trying to train engineers, try to educate employees through on-the- job training.
Evaluation by results only and not by efforts and the process	Since there are few good results, not too many chances of being praised. Morale becomes low.	Separate praise for results and for technological level. Since enthusiasm is important for a project, pay attention even to unfavorable results. Put importance on liberal ideas.
Trying to solve the problem using only familiar factors.	Because new factors were not considered, the problem cannot be solved.	Everything can be a possibility for improvement. Positively utilize all new factors which might be related to improvement.
Insufficient understanding of the objective.	An objective is considered to be fixed and unchanging.	Since an enterprise is mobile, the objective changes. It is common sense that an objective changes because of competition.

Experiments are not well designed. They do not try to forecast experimental results.	Belief that experimentation is an action against the unknown. Therefore, results cannot be forecast.	Ask the question "If this experiment fails,..." and exclude the part which might fail. If it cannot be excluded, include some countermeasures for recovery.
Too much pursuit of theory and mechanism.	Development work delayed.	Make three hypotheses. Review the factors based on these hypotheses and find the most appropriate. It is better not to conduct experiments just for the sake of analysis. When experiments are proposed for analysis, use forecasted conclusions as factors for the experiments.
Frequent changes of the target makes it difficult to focus.	Research becomes inefficient.	For new technology or new product development, a target is a reference. Use a range as the target. Consider the area which is not included in specifications.
Testing cannot be made because equipment is unavailable.	It takes too long to make the equipment.	Use hand-made or batch-testing equipment. Design the equipment so that it may be modified easily. Utilize experimental design techniques and design the equipment based on the knowledge of batch-testing.
No confidence in experimentation.	Experiments are conducted using only the factors which are familiar.	Improve the content of the experiment with various inputs until you have confidence in achieving the objective. Individually set up, reckless experiments lead to failure and confusion.
No consideration of cost and productivity.	Quality is given high priority and it is difficult to utilize manufacturing equipment.	Quality is important, but competitivenss is maintained by function/cost. Cost should be discussed before quality.
Ambiguous experimental attitude	Relying on intangibles for success.	Separate fact and estimate. Check the contents of the experiment and perform the experiment with a team of engineers.

Too many substitutional characteristics	Substitutional characteristic and the real characteristic may not agree. Could cause a large loss at the end.	Rough evaluation is made based on substitutional characteristics. The important evaluation is made with the real characteristic. Accelerated life test is effective for shortening development time.
Ambiguity in the scope of responsibility.	Attempt made to solve the problem on an individual basis.	Discussion must be conducted by a group. Team effort is important.
Concentrating too hard on experimental form rather than thinking in terms of the objective.	Results in failures and activities which are off the target.	It is important to try hard, but there is no wisdom in perspiration. Wisdom precedes perspiration.
Overemphasis on deadlines.	Management does not support engineers because of the deadline.	The role of management is to remove the obstacle to engineers. The lead engineer should act as facilitator. Management should provide the necessary time, support and resources.

Orthogonal Arrays and Linear Graphs:

1. "No." represents the experiment number, and "Column" represents the number of column in the orthognal array.
2. The table of interactions is for finding two-factor interactions between two columns.
3. The classification by groups of the columns of the orthogonal array is indicated by the following symbols in assignment types.

Group	Symbol	Symbols only for $L_{32}(2^{31})$
Group 1	○	Groups 1 and 2
Group 2	◎	Group 3
Group 3	●	Group 4
Group 4	⊙	Group 5

$L_8(2^7)$

No. \ Col.	1	2	3	4	5	6	7
1	1	1	1	1	1	1	1
2	1	1	1	2	2	2	2
3	1	2	2	1	1	2	2
4	1	2	2	2	2	1	1
5	2	1	2	1	2	1	2
6	2	1	2	2	1	2	1
7	2	2	1	1	2	2	1
8	2	2	1	2	1	1	2
Component	a	b	a b	c	a c	b c	a b c
	Grp. 1	Grp. 2		Grp. 3			

2-Column Interaction Table

Column	1	2	3	4	5	6	7
	(1)	3	2	5	4	7	6
		(2)	1	6	7	4	5
			(3)	7	6	5	4
				(4)	1	2	3
					(5)	3	2
						(6)	1
							(7)

(1)

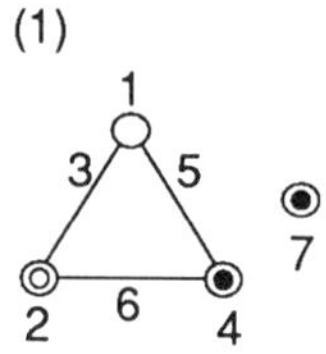

(2)

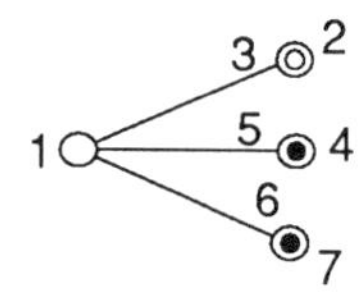

$L_4(2^3)$

(1)

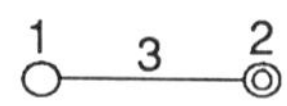

2-Column Interaction Table

No. \ Column	1	2	3
1	1	1	1
2	1	2	2
3	2	1	2
4	2	2	1
Component	a	b	a b
	Group 1	Group 2	

From Orthogonal Arrays and Linear Graphs, Genichi Taguchi, Maruzen (1961) and from Quality Control Manual, Jiro Yamauchi, et al., Japanese Standards Association (1962).

$L_{16}(2^{15})$

Column	1	2	3	4	5	6	7	8	9	10	11	12	13	14	15
1	1	1	1	1	1	1	1	1	1	1	1	1	1	1	1
2	1	1	1	1	1	1	1	2	2	2	2	2	2	2	2
3	1	1	1	2	2	2	2	1	1	1	1	2	2	2	2
4	1	1	1	2	2	2	2	2	2	2	2	1	1	1	1
5	1	2	2	1	1	2	2	1	1	2	2	1	1	2	2
6	1	2	2	1	1	2	2	2	2	1	1	2	2	1	1
7	1	2	2	2	2	1	1	1	1	2	2	2	2	1	1
8	1	2	2	2	2	1	1	2	2	1	1	1	1	2	2
9	2	1	2	1	2	1	2	1	2	1	2	1	2	1	2
10	2	1	2	1	2	1	2	2	1	2	1	2	1	2	1
11	2	1	2	2	1	2	1	1	2	1	2	2	1	2	1
12	2	1	2	2	1	2	1	2	1	2	1	1	2	1	2
13	2	2	1	1	2	2	1	1	2	2	1	1	2	2	1
14	2	2	1	1	2	2	1	2	1	1	2	2	1	1	2
15	2	2	1	2	1	1	2	1	2	2	1	2	1	1	2
16	2	2	1	2	1	1	2	2	1	1	2	1	2	2	1
	a	b	a b	c	a c	b c	a b c	d	a d	b d	a b d	c d	a c d	b c d	a b c d
Component	Group 1	Group 2		Group 3				Group 4							

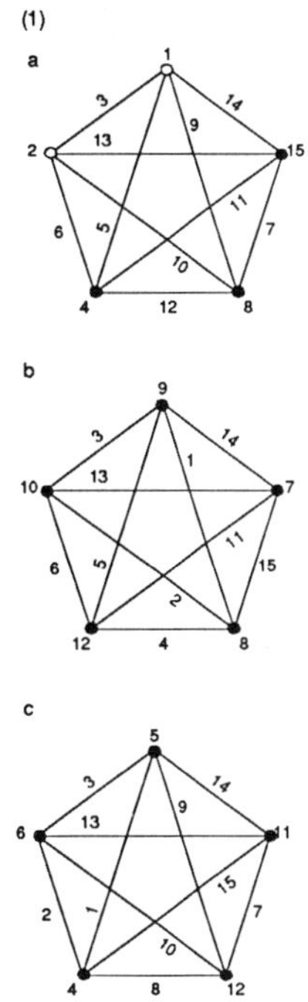

$L_{12}(2^{11})$

Column No.	1	2	3	4	5	6	7	8	9	10	11
1	1	1	1	1	1	1	1	1	1	1	1
2	1	1	1	1	1	2	2	2	2	2	2
3	1	1	2	2	2	1	1	1	2	2	2
4	1	2	1	2	2	1	2	2	1	1	2
5	1	2	2	1	2	2	1	2	1	2	1
6	1	2	2	2	1	2	2	1	2	1	1
7	2	1	2	2	1	1	2	2	1	2	1
8	2	1	2	1	2	2	2	1	1	1	2
9	2	1	1	2	2	2	1	2	2	1	1
10	2	2	2	1	1	1	1	2	2	1	2
11	2	2	1	2	1	2	1	1	1	2	2
12	2	2	1	1	2	1	2	1	2	2	1
	Group 1		Group 2								

Note: There is no linear graph for the $L_{12}(2^{11})$. The interaction component of any two columns is more or less confounded with the remaining nine columns. The sequential analysis method is required to obtain interactions. Do not use this array for experiments with an interaction.

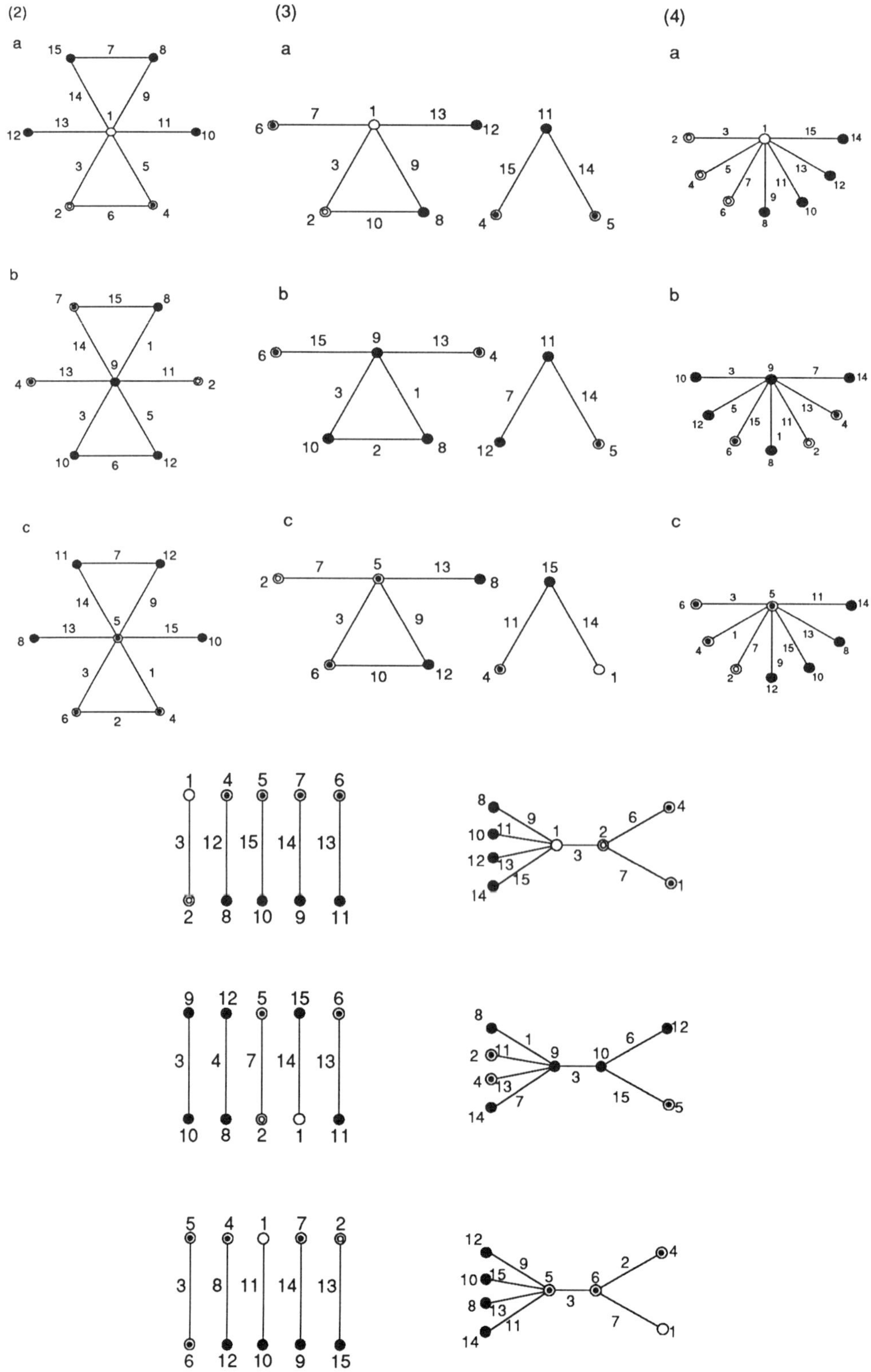
(2)
(3)
(4)
a
b
c

L_{32}

No.	1	2	3	4	5	6	7	8	9	10	11	12	13	14	15	16	17	18	19	20	21	22	23	24	25	26	27	28	29	30	31
1	1	1	1	1	1	1	1	1	1	1	1	1	1	1	1	1	1	1	1	1	1	1	1	1	1	1	1	1	1	1	1
2	1	1	1	1	1	1	1	1	1	1	1	1	1	1	1	2	2	2	2	2	2	2	2	2	2	2	2	2	2	2	2
3	1	1	1	1	1	1	1	2	2	2	2	2	2	2	2	1	1	1	1	1	1	1	1	2	2	2	2	2	2	2	2
4	1	1	1	1	1	1	1	2	2	2	2	2	2	2	2	2	2	2	2	2	2	2	2	1	1	1	1	1	1	1	1
5	1	1	1	2	2	2	2	1	1	1	1	2	2	2	2	1	1	1	1	2	2	2	2	1	1	1	1	2	2	2	2
6	1	1	1	2	2	2	2	1	1	1	1	2	2	2	2	2	2	2	2	1	1	1	1	2	2	2	2	1	1	1	1
7	1	1	1	2	2	2	2	2	2	2	2	1	1	1	1	1	1	1	1	2	2	2	2	2	2	2	2	1	1	1	1
8	1	1	1	2	2	2	2	2	2	2	2	1	1	1	1	2	2	2	2	1	1	1	1	1	1	1	1	2	2	2	2
9	1	2	2	1	1	2	2	1	1	2	2	1	1	2	2	1	1	2	2	1	1	2	2	1	1	2	2	1	1	2	2
10	1	2	2	1	1	2	2	1	1	2	2	1	1	2	2	2	2	1	1	2	2	1	1	2	2	1	1	2	2	1	1
11	1	2	2	1	1	2	2	2	2	1	1	2	2	1	1	1	1	2	2	1	1	2	2	2	2	1	1	2	2	1	1
12	1	2	2	1	1	2	2	2	2	1	1	2	2	1	1	2	2	1	1	2	2	1	1	1	1	2	2	1	1	2	2
13	1	2	2	2	2	1	1	1	1	2	2	2	2	1	1	1	1	2	2	2	2	1	1	1	1	2	2	2	2	1	1
14	1	2	2	2	2	1	1	1	1	2	2	2	2	1	1	2	2	1	1	1	1	2	2	2	2	1	1	1	1	2	2
15	1	2	2	2	2	1	1	2	2	1	1	1	1	2	2	1	1	2	2	2	2	1	1	2	2	1	1	1	1	2	2
16	1	2	2	2	2	1	1	2	2	1	1	1	1	2	2	2	2	1	1	1	1	2	2	2	2	1	1	1	1	2	2
17	2	1	2	1	2	1	2	1	2	1	2	1	2	1	2	1	2	1	2	1	2	1	2	1	2	1	2	1	2	1	2
18	2	1	2	1	2	1	2	1	2	1	2	1	2	1	2	2	1	2	1	2	1	2	1	2	1	2	1	2	1	2	1
19	2	1	2	1	2	1	2	2	1	2	1	2	1	2	1	1	2	1	2	1	2	1	2	2	1	2	1	2	1	2	1
20	2	1	2	1	2	1	2	2	1	2	1	2	1	2	1	2	1	2	1	2	1	2	1	1	2	1	2	1	2	1	2
21	2	1	2	2	1	2	1	1	2	1	2	2	1	2	1	1	2	1	2	2	1	2	1	1	2	1	2	2	1	2	1
22	2	1	2	2	1	2	1	1	2	1	2	2	1	2	1	2	1	2	1	1	2	1	2	2	1	2	1	1	2	1	2
23	2	1	2	2	1	2	1	2	1	2	1	1	2	1	2	1	2	1	2	2	1	2	1	2	1	2	1	1	2	1	2
24	2	1	2	2	1	2	1	2	1	2	1	1	2	1	2	2	1	2	1	1	2	1	2	1	2	1	2	2	1	2	1
25	2	2	1	1	2	2	1	1	2	2	1	1	2	2	1	1	2	2	1	1	2	2	1	1	2	2	1	1	2	2	1
26	2	2	1	1	2	2	1	1	2	2	1	1	2	2	1	2	1	1	2	2	1	1	2	2	1	1	2	2	1	1	2
27	2	2	1	1	2	2	1	2	1	1	2	2	1	1	2	1	2	2	1	1	2	2	1	2	1	1	2	2	1	1	2
28	2	2	1	1	2	2	1	2	1	1	2	2	1	1	2	2	1	1	2	2	1	1	2	1	2	2	1	1	2	2	1
29	2	2	1	2	1	1	2	1	2	2	1	2	1	1	2	1	2	2	1	2	1	1	2	1	2	2	1	2	1	1	2
30	2	2	1	2	1	1	2	1	2	2	1	2	1	1	2	2	1	1	2	1	2	2	1	2	1	1	2	1	2	2	1
31	2	2	1	2	1	1	2	2	1	1	2	1	2	2	1	1	2	2	1	2	1	1	2	2	1	1	2	1	2	2	1
32	2	2	1	2	1	1	2	2	1	1	2	1	2	2		2	1	1	2	1	2	2	1	1	2	2	1	2	1	1	2
	a	b	a	c	a	b	a	d	a	b	a	c	a	b	a	e	a	b	a	c	a	b	a	d	a	b	a	c	a	b	a
			a		c	c	b		d	d	b	d	c	c	b		e	e	b	e	c	c	b	e	d	d	b	d	c	c	b
							c				d		d	d	c				e		e	e	c		e	e	d	e	d	d	c
															d								e				e		e	e	d
																															e
	Grp. 1	Grp. 2		Group 3				Group 4								Group 5															

2-Column Interaction Table for L_{32}

Column \ Column	1	2	3	4	5	6	7	8	9	10	11	12	13	14	15	16	17	18	19	20	21	22	23	24	25	26	27	28	29	30	31
1)		3	2	5	4	7	6	9	8	11	10	13	12	15	14	17	16	19	18	21	20	23	22	25	24	27	26	29	28	31	30
2)			1	6	7	4	5	10	11	8	9	14	15	12	13	18	19	16	17	22	23	20	21	26	27	24	25	30	31	28	29
3)				7	6	5	4	11	10	9	8	15	14	13	12	19	18	17	16	23	22	21	20	27	26	25	24	31	30	29	28
4)					1	2	3	12	13	14	15	8	9	10	11	20	21	22	23	16	17	18	19	28	29	30	31	24	25	26	27
5)						3	2	13	12	15	14	9	8	11	10	21	20	23	22	17	16	19	18	29	28	31	30	25	24	27	26
6)							1	14	15	12	13	10	11	8	9	22	23	20	21	18	19	16	17	30	31	28	29	26	27	24	25
7)								15	14	13	12	11	10	9	8	23	22	21	20	19	18	17	16	31	30	29	28	27	26	25	24
8)									1	2	3	4	5	6	7	24	25	26	27	28	29	30	31	16	17	18	19	20	21	22	23
9)										3	2	5	4	7	6	25	24	27	26	29	28	31	30	17	16	19	18	21	20	23	22
10)											1	6	7	4	5	26	27	24	25	30	31	28	29	18	19	16	17	22	23	20	21
11)												7	6	5	4	27	26	25	24	31	30	29	28	19	18	17	16	23	22	21	20
12)													1	2	3	28	29	30	31	24	25	26	27	20	21	22	23	16	17	18	19
13)														3	2	29	28	31	30	25	24	27	26	21	20	23	22	17	16	19	18
14)															1	30	31	28	29	26	27	24	25	22	23	20	21	18	19	16	17
15)																31	30	29	28	27	26	25	24	23	22	21	20	19	18	17	16
16)																	1	2	3	4	5	6	7	8	9	10	11	12	13	14	15
17)																		3	2	5	4	7	6	9	8	11	10	13	12	15	14
18)																			1	6	7	4	5	10	11	8	9	14	15	12	13
19)																				7	6	5	4	11	10	9	8	15	14	13	12
20)																					1	2	3	12	13	14	15	8	9	10	11
21)																						3	2	13	12	15	14	9	8	11	10
22)																							1	14	15	12	13	10	11	8	9
23)																								15	14	13	12	11	10	9	8
24)																									1	2	3	4	5	6	7
25)																										3	2	5	4	7	6
26)																											1	6	7	4	5
27)																												7	6	5	4
28)																													1	2	3
29)																														3	2
30)																															1

(1)

a

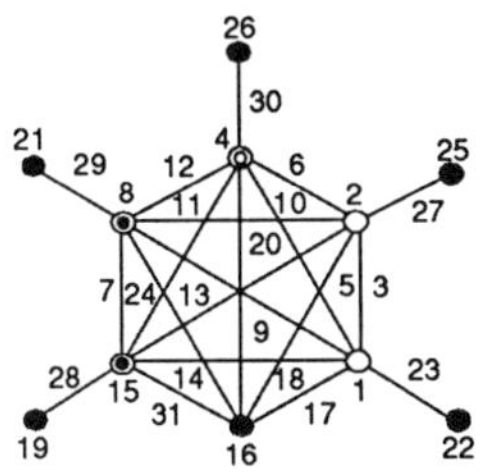

b

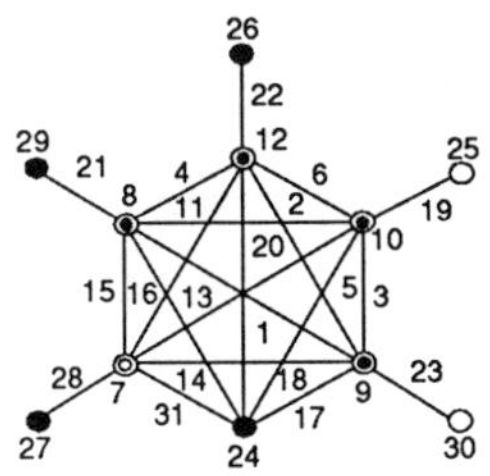

c

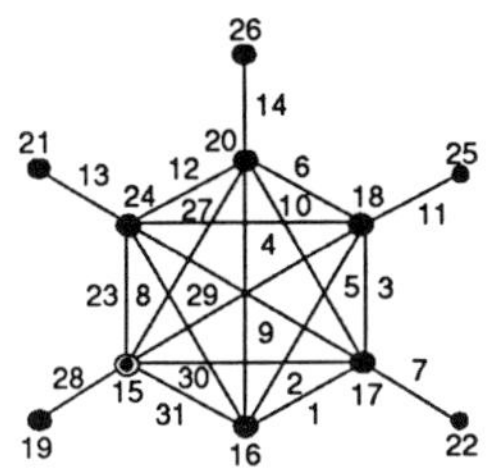

(2)

a

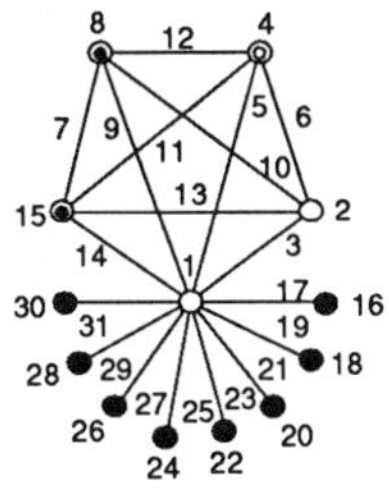

b

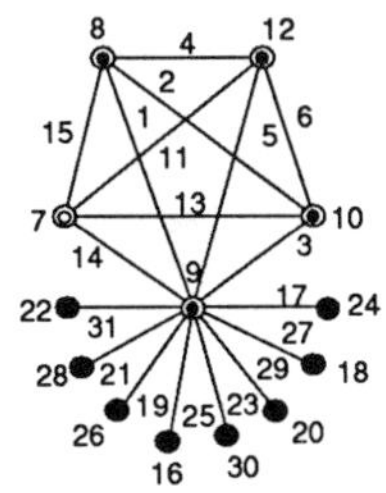

c

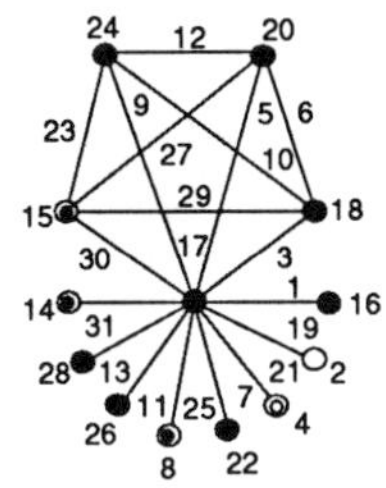

(3)

a

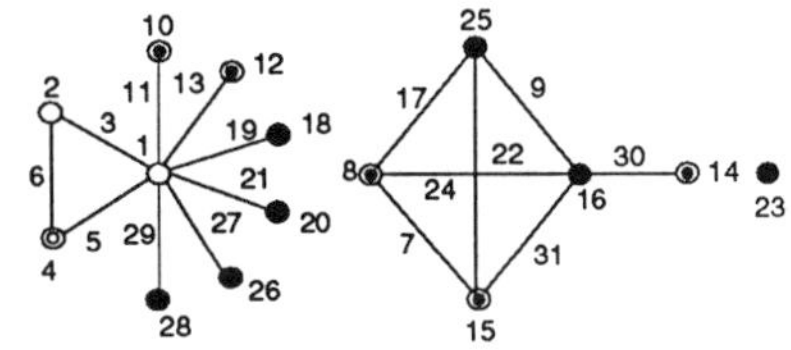

b

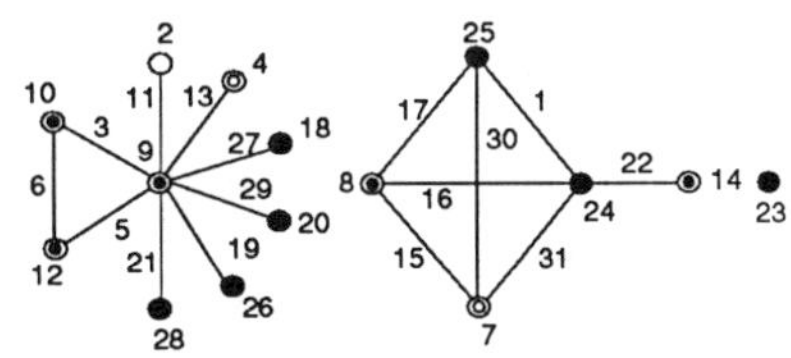

c

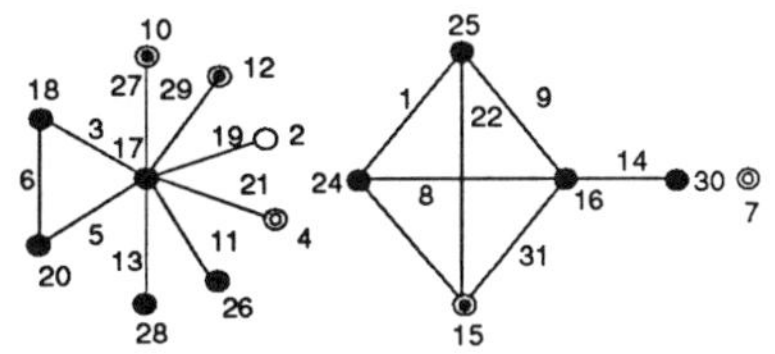

(4)

a

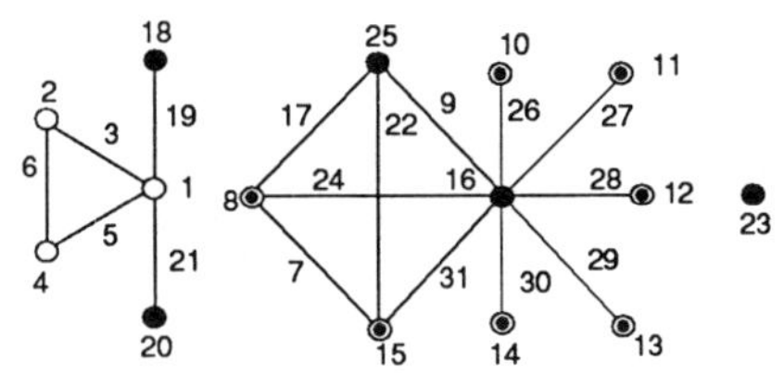

b

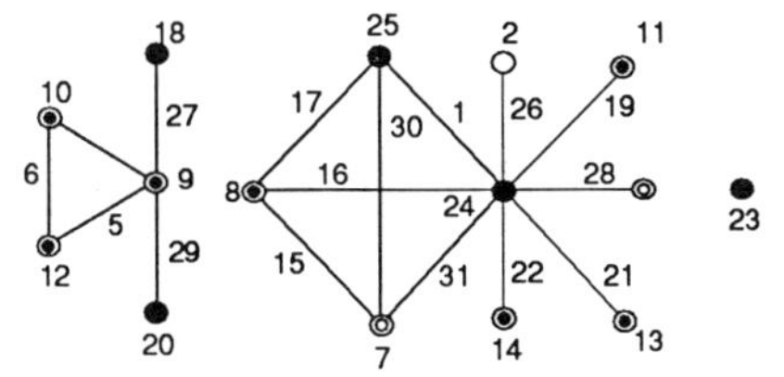

c

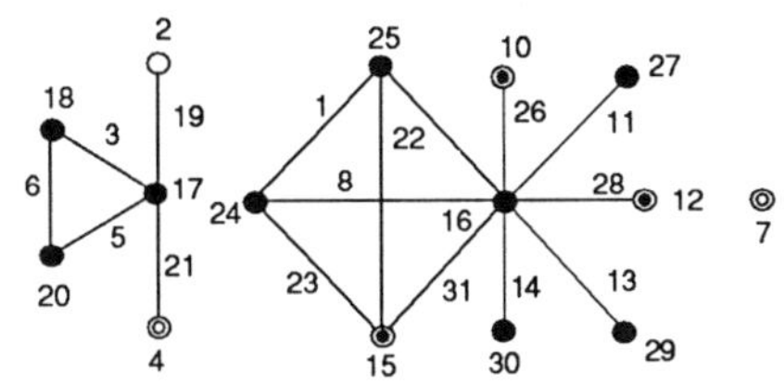

(5)

a

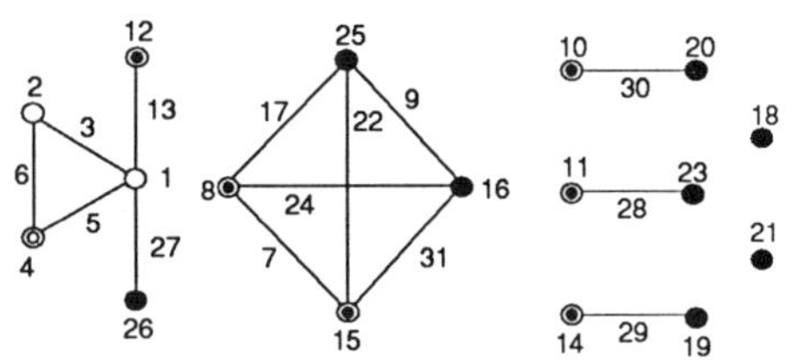

b

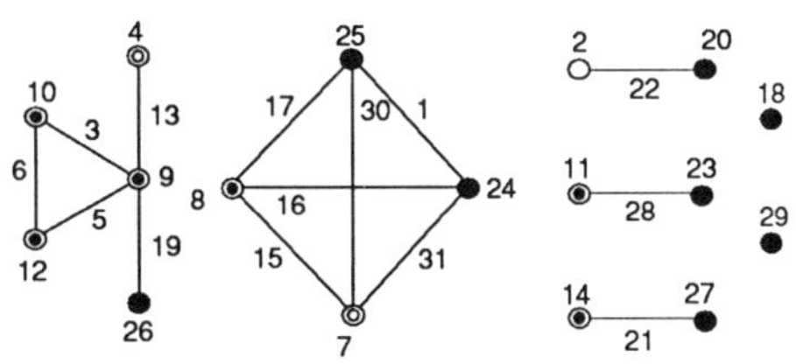

c

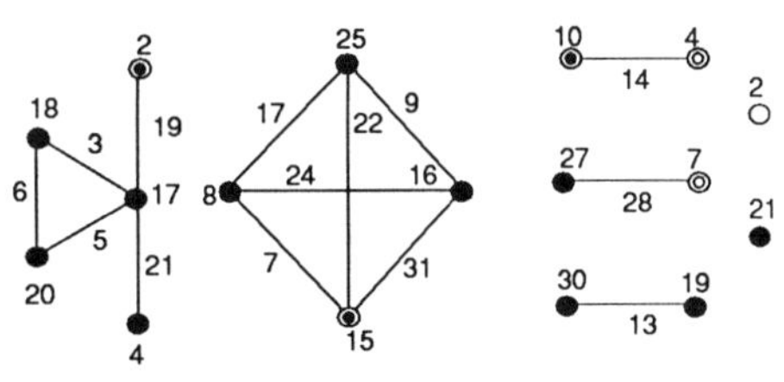

(6)

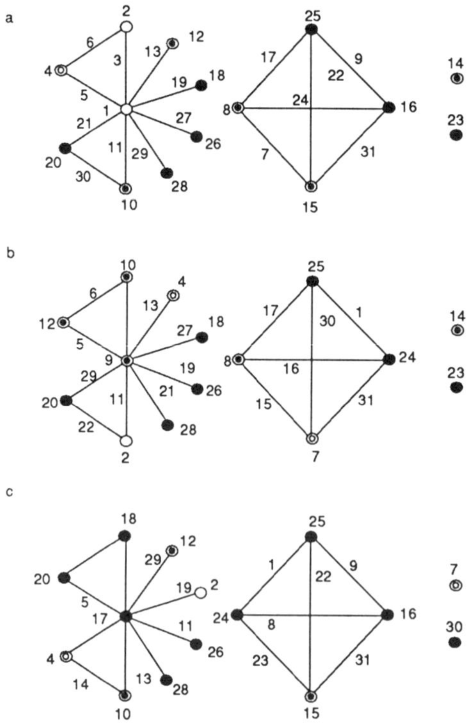
a
2 6 3 13 12 4 19 18 5 1 27 21 26 11 29 20 30 28 10
25 17 9 22 8 24 16 7 31 15
14 23
b
10 6 4 13 12 27 18 5 9 19 29 21 26 20 11 22 28 2
25 17 30 1 8 16 24 15 31 7
14 23
c
18 12 29 20 19 2 5 17 11 26 4 13 14 28 10
25 1 22 9 24 8 16 23 31 15
7 30

(7)

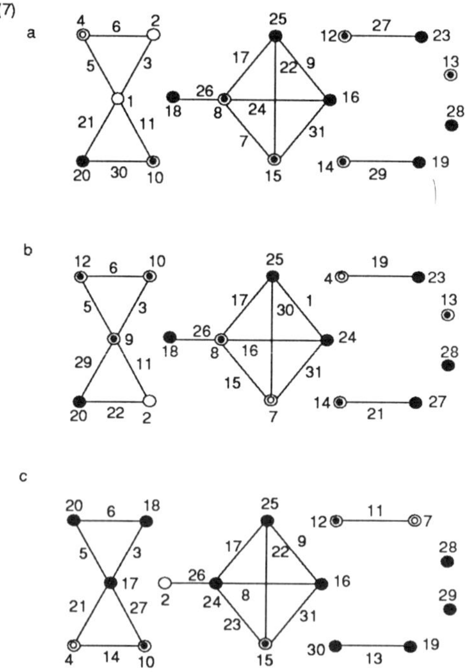
a
4 6 2 5 3 1 21 11 20 30 10
25 17 22 9 18 26 8 24 16 7 31 15
12 27 23 13 28 14 29 19
b
12 6 10 5 3 9 29 11 20 22 2
25 17 30 1 18 26 8 16 24 15 31 7
4 19 23 13 28 14 21 27
c
20 6 18 5 3 17 21 27 4 14 10
25 17 22 9 2 26 24 8 16 23 31 15
12 11 7 28 29 30 13 19

(8)

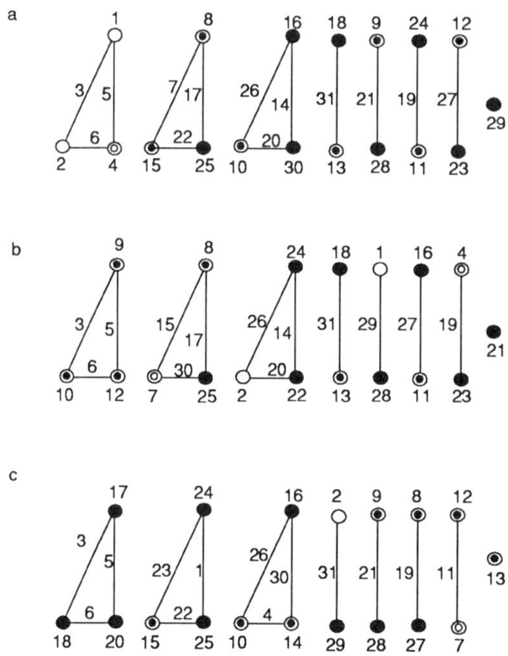
a
1 3 5 6 2 4
8 7 17 22 15 25
16 26 14 20 10 30
18 31 13
9 21 28
24 19 11
12 27 23
29
b
9 3 5 6 10 12
8 15 17 30 7 25
24 26 14 20 2 22
18 31 13
1 29 28
16 27 11
4 19 23
21
c
17 3 5 6 18 20
24 23 1 22 15 25
16 26 30 4 10 14
2 31 29
9 21 28
8 19 27
12 11 7
13

(9)

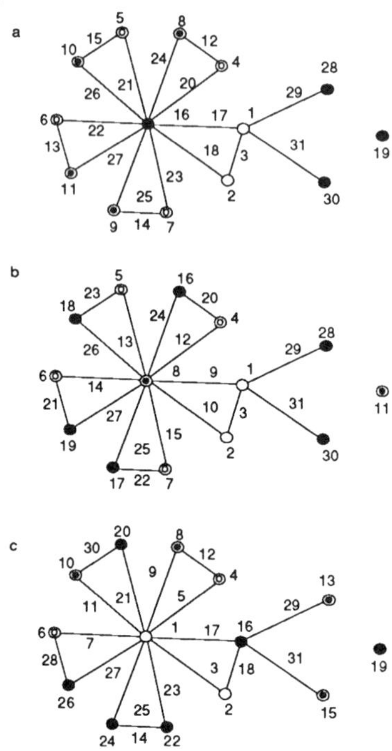
a
5 8 10 15 24 12 4 21 20 26 28 16 17 1 29 6 22 13 27 18 3 31 19 11 23 25 2 30 9 14 7
b
5 16 18 23 20 24 4 13 12 26 28 8 9 1 29 6 14 21 10 3 31 11 27 15 19 25 2 30 17 22 7
c
20 8 30 12 10 9 4 21 5 11 13 1 17 16 29 6 7 28 3 18 31 19 27 23 26 25 2 15 24 14 22

(10)

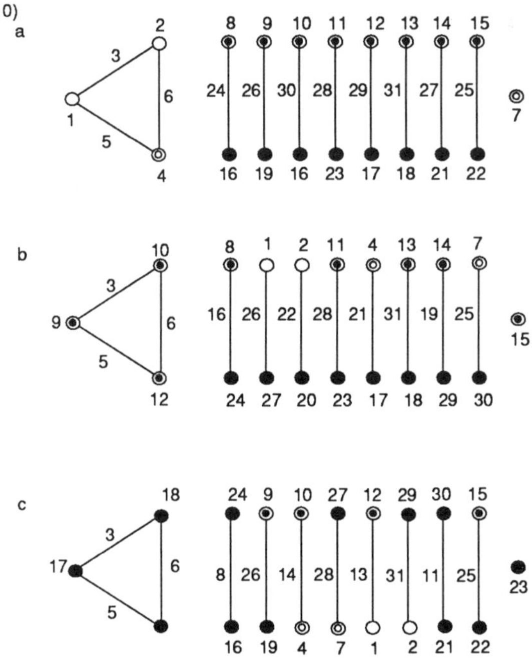

(11)

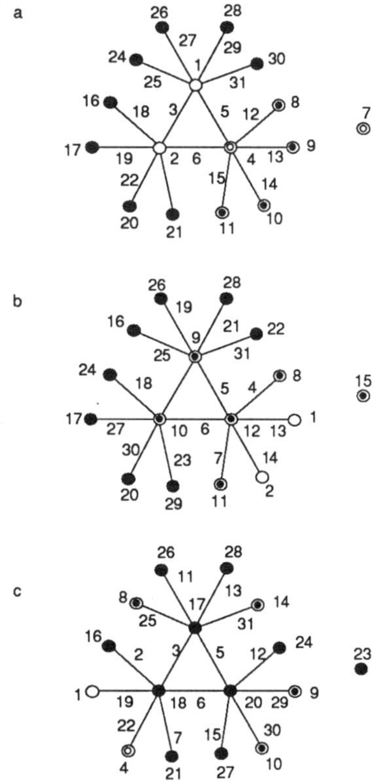

(12)

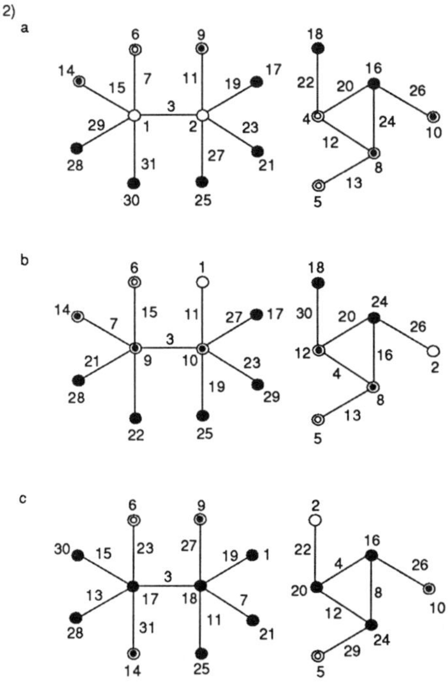

(13)

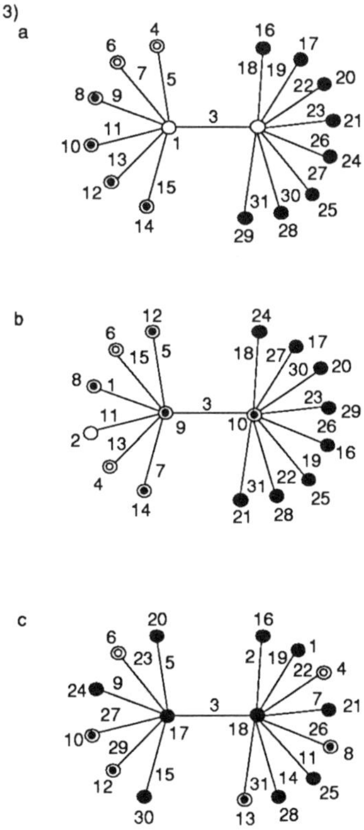

Orthogonal Arrays and Linear Graphs

$L_9\ (3^4)$

No. \ Col.	1	2	3	4
1	1	1	1	1
2	1	2	2	2
3	1	3	3	3
4	2	1	2	3
5	2	2	3	1
6	2	3	1	2
7	3	1	3	2
8	3	2	1	3
9	3	3	2	1
Component	a	b	a b	a b^2
	Group 1	Group 2		

(1)

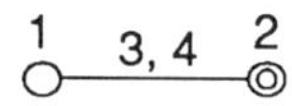

Note: The interaction between columns 1 and 2 takes up two columns.

$L_{18}\ (2^1 \times 3^7)$

No. \ Col.	1	2	3	4	5	6	7	8
1	1	1	1	1	1	1	1	1
2	1	1	2	2	2	2	2	2
3	1	1	3	3	3	3	3	3
4	1	2	1	1	2	2	3	3
5	1	2	2	2	3	3	1	1
6	1	2	3	3	1	1	2	2
7	1	3	1	2	1	3	2	3
8	1	3	2	3	2	1	3	1
9	1	3	3	1	3	2	1	2
10	2	1	1	3	3	2	2	1
11	2	1	2	1	1	3	3	2
12	2	1	3	2	2	1	1	3
13	2	2	1	2	3	1	3	2
14	2	2	2	3	1	2	1	3
15	2	2	3	1	2	3	2	1
16	2	3	1	3	2	3	1	2
17	2	3	2	1	3	1	2	3
18	2	3	3	2	1	2	3	1
	Grp. 1	Grp. 2	Group 3					

(1)

Note: An interaction can be obtained without sacrificing columns. It can be obtained from the binary table for columns 1 and 2. Interactions between 3 columns are partially confounded with the remaining 3-level columns.

$L_{27}(3^{13})$

No. \ Col.	1	2	3	4	5	6	7	8	9	10	11	12	13
1	1	1	1	1	1	1	1	1	1	1	1	1	1
2	1	1	1	1	2	2	2	2	2	2	2	2	2
3	1	1	1	1	3	3	3	3	3	3	3	3	3
4	1	2	2	2	1	1	1	2	2	2	3	3	3
5	1	2	2	2	2	2	2	3	3	3	1	1	1
6	1	2	2	2	3	3	3	1	1	1	2	2	2
7	1	3	3	3	1	1	1	3	3	3	2	2	2
8	1	3	3	3	2	2	2	1	1	1	3	3	3
9	1	3	3	3	3	3	3	2	2	2	1	1	1
10	2	1	2	3	1	2	3	1	2	3	1	2	3
11	2	1	2	3	2	3	1	2	3	1	2	3	1
12	2	1	2	3	3	1	2	3	1	2	3	1	2
13	2	2	3	1	1	2	3	2	3	1	3	1	2
14	2	2	3	1	2	3	1	3	1	2	1	2	3
15	2	2	3	1	3	1	2	1	2	3	2	3	1
16	2	3	1	2	1	2	3	3	1	2	2	3	1
17	2	3	1	2	2	3	1	1	2	3	3	1	2
18	2	3	1	2	3	1	2	2	3	1	1	2	3
19	3	1	3	2	1	3	2	1	3	2	1	3	2
20	3	1	3	2	2	1	3	2	1	3	2	1	3
21	3	1	3	2	3	2	1	3	2	1	3	2	1
22	3	2	1	3	1	3	2	2	1	3	3	2	1
23	3	2	1	3	2	1	3	3	2	1	1	3	2
24	3	2	1	3	3	2	1	1	3	2	2	1	3
25	3	3	2	1	1	3	2	3	2	1	2	1	3
26	3	3	2	1	2	1	3	1	3	2	3	2	1
27	3	3	2	1	3	2	1	2	1	3	1	3	2
Component	a	b	a b	a b^2	c	a c	a c^2	b c	a b c	a b^2 c	b c^2	a b^2 c	a b c^2
	Group 1	Group 2			Group 3								

(1)

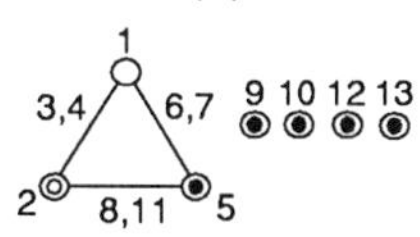

(2)

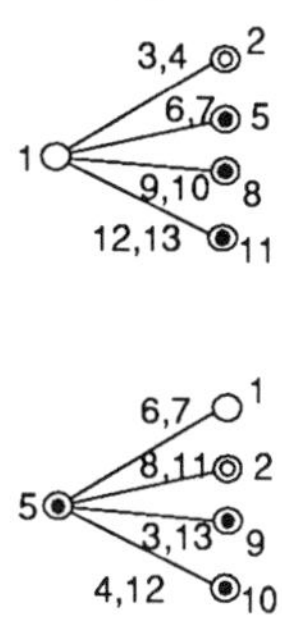

2-Column Interaction Table for L_{27} (3^{13})

Column \ Column	1	2	3	4	5	6	7	8	9	10	11	12	13
	1)	3	2	2	6	5	5	9	8	8	12	11	11
		4	4	3	7	7	6	10	10	9	13	13	12
			1	1	8	9	10	5	6	7	5	6	7
		2)	4	3	11	12	13	11	12	13	8	9	10
				1	9	10	8	7	5	6	6	7	5
			3)	2	13	11	12	12	13	11	10	8	9
					10	8	9	6	7	5	7	5	6
				4)	12	13	11	13	11	12	9	10	8
						1	1	2	3	4	2	4	3
					5)	7	6	11	13	12	8	10	9
							1	4	2	3	3	2	4
						6)	5	13	12	11	10	9	8
								3	4	2	4	3	2
							7)	12	11	13	9	8	10
									1	1	2	3	4
								8)	10	9	5	7	6
										1	4	2	3
									9)	8	7	6	5
											3	4	2
										10)	6	5	7
												1	1
											11)	13	12
													1
												12)	11

Orthogonal Array L_{36}

No. \ Col.	1	2	3	4	5	6	7	8	9	10	11	12	13
1	1	1	1	1	1	1	1	1	1	1	1	1	1
2	2	2	2	2	2	2	2	2	2	2	2	2	1
3	3	3	3	3	3	3	3	3	3	3	3	3	1
4	1	1	1	1	2	2	2	2	3	3	3	3	1
5	2	2	2	2	3	3	3	3	1	1	1	1	1
6	3	3	3	3	1	1	1	1	2	2	2	2	1
7	1	1	2	3	1	2	3	3	1	2	2	3	1
8	2	2	3	1	2	3	1	1	2	3	3	1	1
9	3	3	1	2	3	1	2	2	3	1	1	2	1
10	1	1	3	2	1	3	2	3	2	1	3	2	1
11	2	2	1	3	2	1	3	1	3	2	1	3	1
12	3	3	2	1	3	2	1	2	1	3	2	1	1
13	1	2	3	1	3	2	1	3	3	2	1	2	2
14	2	3	1	2	1	3	2	1	1	3	2	3	2
15	3	1	2	3	2	1	3	2	2	1	3	1	2
16	1	2	3	2	1	1	3	2	3	3	2	1	2
17	2	3	1	3	2	2	1	3	1	1	3	2	2
18	3	1	2	1	3	3	2	1	2	2	1	3	2
19	1	2	1	3	3	3	1	2	2	1	2	3	2
20	2	3	2	1	1	1	2	3	3	2	3	1	2
21	3	1	3	2	2	2	3	1	1	3	1	2	2
22	1	2	2	3	3	1	2	1	1	3	3	2	2
23	2	3	3	1	1	2	3	2	2	1	1	3	2
24	3	1	1	2	2	3	1	3	3	2	2	1	2
25	1	3	2	1	2	3	3	1	3	1	2	2	3
26	2	1	3	2	3	1	1	2	1	2	3	3	3
27	3	2	1	3	1	2	2	3	2	3	1	1	3
28	1	3	2	2	2	1	1	3	2	3	1	3	3
29	2	1	3	3	3	2	2	1	3	1	2	1	3
30	3	2	1	1	1	3	3	2	1	2	3	2	3
31	1	3	3	3	2	3	2	2	1	2	1	1	3
32	2	1	1	1	3	1	3	3	2	3	2	2	3
33	3	2	2	2	1	2	1	1	3	1	3	3	3
34	1	3	1	2	3	2	3	1	2	2	3	1	3
35	2	1	2	3	1	3	1	2	3	3	1	2	3
36	3	2	3	1	2	1	2	3	1	1	2	3	3

Index

D

E

G

I

J

L

M

N

O

P

Q

V